Agriculture under the Common Agricultural Policy

A GEOGRAPHY

TO VAL, JULIE,
MICHAEL AND SARAH

Agriculture under the
Common Agricultural Policy

A GEOGRAPHY

Ian R. Bowler

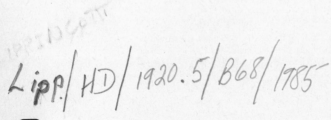

Manchester University Press

Published by Manchester University Press
Oxford Road, Manchester M13 9PL
Copyright © Ian R. Bowler 1985

British Library cataloguing in publication data
Bowler, Ian R.
Agriculture under the Common Agricultural Policy. A geography
1. Agriculture and state. – European
Economic Community countries
I. Title
338.1′81   HD1920.5.Z8

Library of Congress cataloging in publication data

Bowler, Ian R.
  Agriculture under the Common Agricultural Policy.

  1. Agriculture and state—European Economic
Community countries. I. Title.
HD1920.5.B68   1985   338.1′81′094   85–25041

ISBN 0—7190—1095—0 *cased*

Photoset in Great Britain
by Northern Phototypesetting Co, Bolton
Printed in Great Britain
by Unwin Brothers Ltd
The Gresham Press
Old Woking, Surrey
A member of the Martins Printing Group

Contents

Tables and Figures

FIGURES

Preface

Agriculture in all industrial market economies has undergone a transformation during the last three decades. The term 'modernisation' is often used to describe what in effect has become the industrialisation of farming: production characterised by specialisation both on individual farms and whole farming regions, a farm structure of fewer but larger production units, systems of farming that rapidly apply innovations in machinery, plant varieties and artificial fertilisers, marketing arrangements increasingly dominated by co-operatives, food processing firms and supermarket combines.

Governments throughout the world have intervened in these developments of 'modernisation', on the one hand to ameliorate the worst social and economic effects of the changes in matters such as farm incomes and the security of food supplies, and on the other hand to stimulate developments perceived as desirable such as the more rapid evolution of the farm-size structure and the widespread application of new farming technology.

In the last twenty-five years most of Western Europe has been involved in these agricultural changes but within a unique institutional setting: the European Community and its Common Agricultural Policy. Some writers have claimed that with hindsight the EC will be viewed as one of the major events in world history. While this may yet prove to be too fulsome an evaluation, there is no doubt that at present the CAP is the world's most highly developed form of supra-national law and institutional regulations. There is some significance, therefore, in discovering in detail how agriculture has evolved under the CAP and in considering the extent to which the process of farm modernisation has been influenced by the agricultural policies of the Community.

The book has been written mainly with students of geography in mind. Consequently a spatial perspective has been adopted with the changing character of agriculture as its central theme. In separate chapters, for example, attention is given to the changing spatial patterns of agricultural production, food trade, farm size and income, and rural social change. The analysis considers the EC as a whole using data for both individual Members of the Community and regions within those countries.

To fully appreciate the influence of the CAP, however, the reader needs to understand the broad political context of the EC and its policies, together with the methods that have been devised for policy making in the Community. Consequently the early chapters of the book deal with integration and the EC, and with policy making for the development of the CAP. Without these perspectives the character, shortcomings and influence of the CAP cannot be understood. Attention is also given to a description of the main price support mechanisms of the CAP, although briefly since they are treated in detail in many other texts. Necessarily the book takes a broad view of the CAP bringing together material from a wide range of disciplines including political science, economics and sociology as well as geography. Such a broad range of material underlines the complexity of the issues surrounding agricultural policy and it remains the area of Community activity most prone to misunderstanding and misinterpretation. On occasions the misinterpretation seems deliberate by those who oppose the fundamental political ideology of the EC, such as the creation of a supra-national authority. More often, however, the sheer complexity and range of the issues are the cause. Here both published and original material have been simplified and organised with the aim of providing within one book an objective, readily comprehended guide to the nature and consequences of the CAP.

In recent years there has been a growing discontent with the existing provisions of the CAP, not least because of the rising cost of the policy measures and the agricultural surpluses which they appear to encourage. But agreement amongst the Members of the EC on the nature of a revised CAP has proved elusive. Indeed it would be presumptuous of this book to proffer solutions to seemingly intractable problems. Rather the book concludes by clarifying for its readers the central issues that must be resolved for progress to be made on revising the CAP, but always bearing in mind the implications for agriculture and the broader rural society of the Community.

Ian R. Bowler
Leicester
September 1983

1
The CAP and the process of integration in the European Community

The European Community (EC) is a supra-national authority which has increasingly superceded the powers of its Member States.[1] Significantly, agricultural policy has been a major feature of Community integration for historically agricultural issues have tended to lie at the centre of many important world events.[2] The prominence of agricultural policy in the context of a Community that is increasingly industrial and urbanised in character may seem surprising, however, and requires explanation.

'High' politics and the origins of integration
The term 'high politics' is used to describe matters concerned with diplomatic or international strategy, whereas 'low politics' involves issues of economic welfare of a technical or administrative character.[3] This is an important distinction, for the origins of the Community are to be found in matters of 'high politics' even though they were inextricably interwoven with economic considerations. Thus the economic benefits of integration were acknowledged by the founders of the EC, but political unity amongst the nation states of Western Europe was their larger goal. A succession of Frenchmen[4] were principally responsible for perpetuating the concept of European unity from its origins with Duc de Sully in the seventeenth and Abbé Saint-Pierre in the eighteenth centuries. The two men generally credited with the final political impetus of May 1950 for the creation of the Community, however, are Robert Schuman – then the French Foreign Minister – and Jean Monnet – then head of the French Commissariat General du Plan. It was Schuman who advanced the proposal to place the coal and steel resources of Western Europe under a joint authority, with a view to ending the historic rivalry between the nation states and effectively preventing further military conflict. The 'Schuman Plan', as it came to be

called, emerged in a climate of international opinion which, after the ravages of two world wars, sought reconciliation and co-operation between nations in the rebuilding of Western Europe. A customs union, for example, had been established in 1948 by Belgium, the Netherlands and Luxembourg (Benelux), while in 1949 a Council of Europe had been created with a Committee of Ministers and a Consultative Assembly of delegates from the national parliaments, albeit with purely advisory functions. By 1950 permanent agreements had also been reached on trade (General Agreement on Tariffs and Trade (1947) – GATT), economic co-operation (Organisation for European Economic Co-operation (1948) – OEEC), and defence (North Atlantic Treaty Organisation (1949) – NATO). Nor was closer economic and political co-operation a prerogative of Western Europe as witnessed by the formation in 1949 of the Council for Mutual Economic Assistance (COMECON) amongst Communist states. Indeed with the passage of time a wide variety of economic and political groupings has developed so that the EC can claim only to have provided, by a 'demonstration effect', an alternative form of organisation for international relations. The 'Schuman Plan' flourished in these conditions and led, in April 1951, to the Treaty of Paris which established the European Coal and Steel Community (ECSC) encompassing France, West Germany,[5] Italy and the three Benelux countries.

Another concern of 'high politics' emerged in the 1950s to further the process of international co-operation and, in the case of the EC, to turn that process into one of integration. This distinction between co-operation and integration is also an important one for integration is distinguished by the delegation by participants of a portion of their national sovereignty to a body with supra-national powers.[6] The threat of Communist expansion and the 'Cold War' of the 1950s gave integration its political momentum – a trend amongst the democratic nations of Western Europe encouraged at the time by the United States. At one level of argument, therefore, the EC can be interpreted as a response to the challenge issuing from the international political divisions and alignments of the Second World War. The formation of a 'power-block' by West European countries, in this perspective, was to counteract the influence of the United States, the Soviet Union, China, and Japan: to create a pentagonal rather than bipolar world.[7] In practice the EC has operated as an economic rather than military 'super-power' but it is no accident that the enlargement of the Community, including prospective members, has been largely determined by membership of NATO. Nor can the political vacuum caused by the loss of overseas empires by many

European powers be discounted in an explanation of the development of the Community.[8] Most West European countries have had to reorientate their political and economic relations in recent decades, and have sought security and mutual benefit from a closer co-operation largely absent during Europe's colonial area. Indeed, this argument has been extended to suggest that the Community was formed to strengthen and perpetuate the capitalist system of social and economic organisation in Western Europe. From a Marxist perspective, therefore, this 'imperialist integration' is a partial reform supported by the bourgeoisie to keep the masses under control.[9]

While the origins of integration are open to a variety of interpretations, the specific events leading to the formation of the EC are undisputed and well known. An initiative came from a small élite, supported by informed public opinion rather than the mass of the population, which found expression in a meeting of the Foreign Ministers of the ECSC at Messina in June 1955. A committee of experts presided over by Paul-Henri Spaak, the Belgian Foreign Minister, was set up to examine the problems of establishing a common market wider than the ECSC. The outcome of the work of this committee, and of subsequent negotiations, was the Rome Treaty of March 1957 signed by Belgium, France, West Germany, Italy, Luxembourg, and the Netherlands (The Six). The Treaty established the European Economic Community, together with an agreement on setting up the European Atomic Energy Community (Euratom), with institutions which were in some respects similar to those of the ECSC. Three existing institutions of the ECSC were absorbed and given new duties and powers – the Assembly, the Council of Ministers and the Court of Justice – but the European Commission and Permanent Representatives (Ambassadors) to the Community were important new additions to the institutional structure of the Community.

Since the formation of the EC, the momentum towards political unity has been largely dissipated, yet issues of 'high politics' have maintained the need for further development of the Community. For example, on the world scene the confrontation between capitalism and communism has deepened in recent years, while the increasing economic problems of inflation, unemployment, and resource scarcity have developed international as well as national dimensions and have called for co-operation between nation states. More particularly, the EC has become the major customer and a main supplier of agricultural trade in the world economy, accounting for nearly eighteen per cent of the total value of that trade. Events in the Community, therefore, have a significance for all other nations in the world. In international trade, for example, the implications of increased agricultural production in the EC are equally important whether the extra output is consumed within the Community, so replacing alternative supplies, sold to other countries where it may displace the products of traditional suppliers, or given as food aid to developing countries. It is often argued, for example, that the policies of agricultural price stability, export and import control within the EC are at the expense of increased price instability for the world economy in general. When agricultural trade provides a major source of export revenue, as it does for many developing countries, the agricultural policy of the Community assumes an international rather than a purely domestic significance.

Agricultural policy in the EC, therefore, must be judged in the context of a Community in which matters of 'high politics' often override narrowly defined economic considerations. The periodic crises faced by the Common Agricultural Policy (CAP) in budgetary, legislative, and policy matters may be resolved not on the grounds of the intrinsic merits of the Policy, but because of the need from an international viewpoint to maintain the EC as a viable political entity.

The process of integration
There is little in the origins of the EC to presage the importance achieved by the CAP in later years. For an explanation we must turn to another issue with political overtones, namely the nature of the integration process. There remains debate on whether 'integration' should be viewed as a process or the condition of unification,[10] and for simplicity the term is used here in both senses. Broadly three different schools of thought can be identified, but with the passage of time a wide variety of interpretations of 'integration' have been advanced by political scientists.

The 'federalist' school views the process of international integration mainly in political terms from the premise that the nation state can no longer provide its people with military security and economic prosperity. The nation state is interpreted as a temporary, transitory, and now anachronistic form of territorial organisation,[11] with federalism offering a method of obtaining political union among separate states and an alternative to the nationalism inherent in the nation state structure of Western Europe. A federal structure divides powers of government between two levels of government – between central and independent national authorities. In fact, a number of federal structures are possible, and it can be argued that a regional devolution of powers is taking place spontaneously within most large, complex nation states. Further, a federal structure is held to accommodate two divergent pressures: on the one hand pressure to create large units to compete in the world political and economic arena; on the other hand pressure for regional autonomy and identity as exemplified by the independence movements in Scotland, Brittany, and the Basque region of Spain. Some observers[12] have gone so far as to argue that the success of a supra-national authority will depend on the revitalisation of local government as a counterbalance to the remoteness from local issues of a centralised bureaucracy.

A federalist analysis of the integration process, therefore, tends not to take the nation state as the preferred political structure. Rather it seeks to clarify the tasks of government and administration in the modern world and to argue about the size of unit in which these tasks can be most efficiently carried out. The influence exerted by federalist arguments, however, has waxed and waned in Western Europe over the past three decades. Resistance organisations in the Second World War identified strongly with federalism, and the Cold War of the 1950s perpetuated its appeal. Indeed at one time federalism was heralded as a 'revolutionary' response to the problems of conflict that plagued Europe in the first half of the twentieth century. But by the 1960s its momentum had been lost. The rejection by the French National Assembly of the European Defence Community in August 1954 presaged the waning influence of federalist arguments, although national views diverged on policy areas other than defence.[13] In addition, the EC has operated successfully at an economic level without the need for political developments, and formal political institutions have not been required for many political decisions. In sum, many observers believe that the Community has stabilised at the highest level of political and economic co-operation possible among sovereign states, while more recent changes in international relations have raised doubts on the relevance of the Community, and its further evolution, as a framework for dealing with many of the contemporary problems facing its members.[14] Federalist arguments and analyses, however, are not completely redundant. They have been used recently, for example, to suggest a framework for resolving the problems of the CAP.[15]

The 'functionalist' school views the process of integration not in these legal–political terms but rather as an emerging community of economic or materialistic interest: frontiers are made meaningless through the continuous development of common interests across them. Unlike federalists, functionalists hold no firm view of the final political outcome. At a simple level, an automatic process of integration is held to operate whereby integration in one sector of the economy has a 'spill-over' effect in causing necessary integration in other sectors, and technocrats rather than politicians hold the key role in the process. Experience in the EC, however, has shown this mono-causal theory, sometimes called 'the cumulative logic of integration' and almost deterministic in its projected outcome, to be too simplistic largely because political and welfare issues cannot be separated in real world conditions. Opposition has been raised, for example, to the reallocation of welfare spending through a strongly developed Community regional policy, while institutional and political factors have been a greater impediment to integration than once thought. Thus a closer analysis suggests that a 'quantitive leap'[16] was involved in the formation of the Community, and not just an inexorable and inevitable process of integration set in motion by the ECSC. Another reason for the absence of automatic 'spill-over' is the previously described division made by national governments between issues of 'high' and 'low politics'. Matters that are considered as 'low politics', such as policy harmonisation, are amenable to 'spill-over'. But issues of 'high politics' concerned with diplomatic or international strategy are not open to integration by this process; they require separate political initiatives. Monetary union, for example, has fallen into the latter rather than the former group with adverse consequences for the development and effective operation of the CAP. Only political initiatives by the French and West German governments in the autumn of 1978 produced the necessary momentum for establishing the European Monetary System (EMS).

Thus a 'neo-functionalist' school has come to dominate the literature.[17] Its adherents deny any automatic or pre-determined process of integration but argue the need for political activity and co-operation as well as functional integration in building up centralising institutions. Political choices have to be made and neo-functionalists take a pluralist view of political power. This assumes the existence of organised pressure groups for lobbying governments, but with competition between groups ensuring balanced, democratic decisions. Another precondition is the growing bureaucratisation of decision making in Member States leading to the development of central institutions.[18] Both the bureaucratic and political élite groups in these central national institutions, with their loyalties transferred to the supra-national institutions, act as the driving force of functional integration. A convergence of demands within and among the Member States leads the élite groups to promote further integration and provide solutions to internal conflicts.

Alternative interpretations to, and elaborations of the federalist, functionalist, and neo-functionalist schools of thought have been provided by a number of writers as the shortcomings of the earlier theories have become evident. 'Fragmented issue-linkage', for example, has been advanced by Haas.[19] Summarising, the policies pursued by the EC are interpreted as dealing more with 'turbulence' in certain sectors of the world economy than with achieving regional political integration. This viewpoint goes some way to meeting the main criticism of earlier theories that they advance programmatic and normative approaches to the building of a united Europe.[20] Arguably they have diverted attention from the actual towards the ideologically necessary dynamics of a multi-national organisation, with a consequent underestimation of the persistence and viability of the nation state and its identification with sovereignty. In practice the nation state has exerted a resistance to closer integration through the CAP and other policy sectors of the economy, to the extent that governments are often seen as merely aggregating national domestic positions at the international level. Indeed, pessimism about the extent and speed with which closer economic and political union can be achieved has been voiced within the Community by the Marjolin Report.[21] However, the resistance of Member governments to transferring sovereignty to the EC is only in part a manifestation of nationalism. More importantly, Community institutions have not shown they are capable of guaranteeing internal political stability and concensus. Efforts to establish a closer economic or political union would tend to undermine the authority of national governments without an adequate alternative being created. With this view the failure to transfer authority can be justified as a responsible, rational stance.

There is a second area of divergence between the theory and practice of integration in the EC. Most politicians have not transferred their loyalties to the Community nor allowed decision making to become dominantly managerial or technocratic in nature. Certainly the political élite groups played a prominent role ('élite-pull') in the creation of the EC, but in the subsequent evolution of the Community a measure of popular support ('permissive concensus') has been necessary. In democracies, political élite groups remain accountable to their electorates or party members, and while the views of an élite can provide a lead, they cannot be so progressive as to alienate support for that élite group. Thus national politicians have had to respond to domestic political pressures and be seen defending vital national interests, however defined, at the EC level. The distinction foreseen by neo-functionalists between politics and economics has never materialised, therefore, and politicians rather than technocrats have controlled the integration process.

Moreover electorates have been unwilling to pressure their governments into transferring powers to a supra-national body and its bureaucracy. For example, the relatively low 'turnout' at the 1979 elections for the European Parliament has been interpreted as demonstrating a lack of widespread support for the EC (Table 2); and a number of recent opinion polls within the Community indicate a static or falling level of support for supra-national institutions. Most politicians, therefore, have retained a divided loyalty, viewing the national political system as their primary concern with the Community secondary in their calculations. At present the EC neither has sufficient power nor offers a viable alternative to existing national political institutions.

Alternative views of the integration process in the EC impinge upon the CAP. Different viewpoints can be adopted by individuals, political parties and pressure groups, and these viewpoints influence attitudes in decision making for sectors of the economy such as agriculture. A functionalist or neo-functionalist view has been particularly influential, largely replacing the initially important federalist view of integration in the Community. Thus the development of the CAP appears to have been pursued in part with a 'spill-over' effect in mind, and has been defended in certain quarters because of its catalytic effect on the wider process of integration rather than for the intrinsic qualities of the policy measures. Examples of 'spill-over' from the CAP include political precedents set by agricultural policy negotiations, the co-ordination of structural policies and the supporting section of the Agriculture Fund (Fonds Europeén d'Oriéntation et de Garantie Agricole – FEOGA), a feature not required by the Treaty of Rome, the development of regional policy, and closer monetary union. Federalists, however, hold a more ambivalent view of economic policy since integration for them is more a political than economic process. Their principal concern lies in the allocation of decision-making powers between supra-national and regional authorities.

Types of integration
The European Community was established as a customs union in July 1958 although some internal barriers restrict movement and trade within the Community to the present day. The Community subsequently evolved into a common market with the agricultural sector alone developing towards full economic union. This statement implies not only that there is a spectrum of

integration forms[22] but that the EC continues to alter its position on that spectrum. Indeed it has been suggested that because the pace of integration varies with different functions in the Community, a new type of conjunction of states is emerging which has no model in the past and has not yet taken on a final shape.

Six forms of integration, however, are generally recognised. The most simple form of integration is termed 'proximity integration' and is exemplified by the inter-relationship of the Canadian and United States' economies. A free trade area, by comparison, eliminates tariffs on trade between members but has no common external tariffs. Indeed the European Free Trade Association (EFTA) was the form of integration selected by the United Kingdom and six other countries in 1959 as the alternative to joining the EC.[23] A customs union is a more elaborate form of integration requiring the elimination of all internal tariffs, as in a free trade area, the creation of a common external tariff on foreign trade with non-participating countries, and the apportionment of customs revenue among the Member States according to an agreed formula. A new strand of economic theory has been developed on customs unions stemming mainly from the pioneering work of Viner in the early 1950s.[24] The theoretical work distinguishes between trade creation and trade diversion. Trade creation increases welfare by replacing the consumption of higher-cost domestic products by lower-cost products of a partner country, so leading to an improvement in resource utilisation. Trade diversion, by contrast, reduces welfare by replacing lower-cost imports from external sources by higher-cost production from within the customs union. The benefits of a customs union, therefore, can be seen as a balance between these two effects, although later writers[25] have concluded that the dynamic gains from a change in market size can be more important than the static gains or losses due to trade creation and diversion. Consequently, the economic advantages claimed for members of a customs union are economies of large-scale production made possible within an enlarged market area, an increase in specialisation with improved productivity, more efficient allocation of economic resources, lower production costs, and an improved competitive position relative to third parties leading to an improved level of economic welfare for members of the union. Taking a long-term view, the balance of theoretical argument has been overwhelmingly in the favour of a customs union.[26]

A common market develops from a customs union with the unrestricted movement of the factors of production such as labour, finance capital and entrepreneurship. Indirect and direct taxation of goods, services, and profits have to be aligned to create neutrality of internal competition, a process termed 'negative' integration.[27] 'Positive' integration by comparison requires the formation of a common currency and harmonised monetary, fiscal, and social policies, leading to a higher form of integration termed 'economic union'. The co-ordination and harmonisation of economic policy on matters such as public expenditure and public purchasing, however, implies the relegation of national agencies to an executive rather than policy-making role. This alteration in the political structure of Member countries is a decisive distinction between a common market and an economic union, and is the main reason why full economic union has not been achieved by any integration scheme to date. Political union is the ultimate form of integration but requires a political will without prospect amongst the nation states of Western Europe. Integration, therefore, is a process of political as well as economic transformation implying control through some form of new central institution, what has been termed 'the response to the challenge of modernity'.[27]

The significance of agriculture

More progress has been made towards economic union in agriculture than in any other sector of the economy and consequently the CAP is often viewed as the cornerstone of the EC. In part, this stems from the involvement of agricultural policy with 'high politics' and the role of the CAP in creating a 'spill-over' effect in the development of the EC. There are, however, a number of more specific reasons for the elevation of agriculture to the forefront of the movement towards economic union.

First, the need to preserve a food supply base has been one of the most fundamental lessons learned by Europe from past wars,[29] a lesson reinforced by the pessimistic projections of world population in relation to global food supplies.[30] Consequently, a 'siege mentality' gave the expansion of agriculture, together with a high level of self-sufficiency, as major national policy goals in the two decades following the Second World War. Although international conflict now seems more probable in economic rather than military spheres, the strategic argument remains to give prominence to the agricultural sector in the context of the EC.

Secondly, a consensus of opinion existed in the late 1950s and early 1960s that there must be a Common Agricultural Policy. Agriculture, through its influence on food prices, exerts a major influence on living costs and hence on wage levels and industrial costs. Consequently, agriculture was regarded as a central area for co-operation if progress in the industrial field was to be ensured and if a customs union was to function effectively. In view of the protectionist tradition of most West European countries, it is not surprising that the original members of the Community chose a similar system for themselves. Indeed, the removal of the protective nature of their existing policies was deemed neither politically acceptable nor socially desirable. Alternative systems would have required too big a break with tradition and a departure into mechanisms of policies in which Member countries were for the most part inexperienced, and for which the necessary bureaucratic structure did not exist. Moreover, no Member State found it necessary to defend the *status quo* against the formation of a CAP. Some governments were pleased to pass on to a higher authority the troublesome nature of agricultural policy, while others genuinely believed that the agricultural problems of Western Europe could be more readily resolved at a supra-national rather than national level. Events have proved this view to be incorrect, but it nevertheless contributed to the initial support for a CAP.

Thirdly, also involved in the moulding of the Treaty of Rome was the resolution of conflicting national economic interests, especially between France and West Germany. A balanced distribution of opportunities and benefits was required between these two States. The latter country had clear prospects of economic gain from industrial integration and the freer movement of industrial goods and services within the EC. But France, less strongly placed in the industrial sector, required compensating gains for her agricultural industry and in particular markets for her rising surpluses of agricultural products. In a sense, therefore, the CAP represented a balancing of national interests: the pressure for industrial integration forcing through the acceptance of a common policy for agriculture.

The CAP also acknowledged the political power of the 'agricultural vote' at the time when the EC was established. Over a quarter of the working populations of France, Italy and Luxembourg were engaged in agriculture, so Member governments required the acquiescence, if not the agreement of the agricultural sector in order to gain general domestic political support for accession to the Rome Treaty. A common policy favourable to the agricultural industry ensured the necessary political support for the EC.

Finally, the CAP has retained its position as the major claimant on the Budget of the Community, although it must be stressed that this results more from the lack of progress in other fields of economic integration than from the nature of the CAP itself. For as long as agriculture absorbs a high proportion of the Community's finances, and the administrative and legislative time of its institutions, so the industry will retain its contemporary significance in the life of the Community.

Conclusion

Significantly, political union is rarely explicit in the contractual agreements of parties to economic integration, and the EC is no exception. Such agreements, nevertheless, are politically inspired and political considerations underlie many of the issues which are fundamental to the role of the CAP in the Community. To a considerable extent, therefore, the CAP is inextricably caught up with issues of 'high politics', global political considerations, and conflicting views of what the EC is or should be.

Taken together, the arguments deployed in this chapter explain why more progress has been made towards economic union in agriculture than in any other sector of the economy, and why the CAP is often viewed as the cornerstone of the EC, especially by the European Commission. Unfortunately, the degree of success of the Community is now often judged by the performance of the CAP, and the 'cornerstone' view of agricultural policy acts as a major constraint on its future development. Any devolution of responsibility for policy to Member States, for example, is interpreted as a weakening of the Community's supra-national character and a threat to its long-term survival. The fear is that a devolution of responsibility could lead to restrictions in the free trade of farm products, which could then spread to industrial products. Such restrictions might then impinge on the political field and upset the basic relationships between the Member countries. But it could well be that the resolution of many of the problems facing the CAP at present requires some devolution of responsibility for agricultural policy. With this viewpoint, any expansion in other areas of policy within the Community, with a commensurate reduction in the emphasis given to the 'cornerstone' role of the CAP, would assist the evolution of a more successful agricultural policy.

Policy formulation, therefore, is not merely an agricultural or even an economic matter, and to judge the CAP purely in these terms is to ignore the wider political realities of the EC. In particular, vested national interests retain a primacy in the affairs of the Community, even if those interests are defined both in subjective and fluid ways by each Member State.[31]

Notes

1 Legally there are three Communities – the European Coal and Steel Community (ECSC), the European Economic Community (EEC), and the European Atomic

Energy Community (Euratom) – but in July 1967, following the 'Merger Treaty' of April 1965, their principal institutions were combined. For convenience the widely used terms 'Common Market' and 'European Community' (EC) are used throughout this book to refer to the grouping of Member States both before and after 1967.

2　P. M. Raup, 'Constraints and potentialities in agriculture', in R. H. Beck *et al*, *The changing structure of Europe: economic, social and political trends*, (Oxford, Oxford University Press, 1970), pp. 129–30.

3　R. P. Morgan, *High politics, low politics: towards a foreign policy for Western Europe*, The Washington Papers 11, (London, Sage, 1973), p. 8.

4　H. A. Schmitt, 'French politicians and the European Communities: the record of the 1950s', in S. N. Fisher (ed.), *France and the European Community* (Columbus, Ohio State University Press, 1964), pp. 59–81.

5　In this book the term 'West Germany' refers to the Federal Republic of Germany and 'Ireland' to the Republic of Ireland.

6　F. R. Willis (ed.), *European integration* New York, (New Viewpoints, 1975), p. 53.

7　G. Mally, *The European Community in perspective* (Massachusetts, Lexington, 1973).

8　M. Blacksell, *Post-war Europe: a political geography* (Folkestone, Dawson, 1977), p. 83.

9　L. Afanas'ev and V. Kolovnyakov, *Contradictions of agrarian integration in the Common Market* (Moscow, Progress, 1976).

10　B. Balassa (ed.), *The theory of economic integration* (London, George, Allen and Unwin, 1962), p. 1.

11　Mally, *The European Community*, p. 4.

12　P. M. Raup, 'Constraints and potentialities in agriculture', in Beck, *The changing structure of Europe*, pp. 126–70.

13　R. Hansen, 'Regional integration: reflections on a decade of theoretical efforts', in M. Hodges (ed.), *European integration: selected readings*, (Harmondsworth, Penguin, 1972), pp. 184–200.

14　S. J. Warnecke, *The European Community* (New York, Council for European Studies, 1978).

15　J. Marsh, 'European agricultural policy: a federalist solution', in B. Burrows *et al*, *Federal solutions to European issues* (London, Macmillan, 1978), pp. 147–60.

16　L. N. Lindberg and S. A. Scheingold, *Europe's would-be polity* (New Jersey, Prentice Hall, 1970).

17　R. J. Harrison, *Europe in question: theories of regional international integration* (London, George, Allen and Unwin, 1974), pp. 75–92.

18　H. Wallace *et al* (eds.), *Policy making in the European Communities* (London, John Wiley, 1977).

19　E. B. Haas, *The obsolescence of regional integration theory* (Berkeley, Institute of International Studies, 1975).

20 Warnecke, *The European Community*, p. 5.
21 Commission of the European Communities, *Report of the Study Group on 'Economic and Monetary Union 1980'* (Brussels, 1975).
22 D. Swann, *The economics of the Common Market* (Harmondsworth, Penguin, 1972), pp. 29–31.
23 EFTA was formed in November 1959 by the signing of the Stockholm Convention by Austria, Denmark, Norway, Portugal, Sweden, Switzerland, and the United Kingdom. It became effective in July 1960.
24 J. Viner, *The customs union issue* (New York, Carnegie Endowment for International Peace, 1950).
25 P. Robson, *The economics of international integration* (London, George, Allen and Unwin, 1980).
26 B. Bracewell-Milnes, *Economic integration in East and West* (London, Croom Helm, 1976).
27 J. Pinder, 'Positive integration and negative integration: some problems of economic union in the EEC', in Willis, *European integration*, p. 53.
28 Harrison, *Europe in question*, pp. 99–102.
29 Raup, in Beck, *The changing structure*, p. 126.
30 G. Allen, 'Agricultural policies in the shadow of Malthus', *Lloyds Bank Review* CXVII (1975), pp. 14–31.
31 Wallace, *Policy making*, p. 303.

2
Policy-making for agriculture: the Central Institutions

Agricultural policy was a troublesome and contentious issue in most West European countries long before the EC was formed. In some respects, therefore, the Community has merely aggregated and so magnified the long-term problems of agriculture in the Member States. In other respects, however, the EC has created new difficulties for the development of sound agricultural policies. In particular, the process of policy making in the central institutions of the Community has inhibited the development of a rational, flexible and equitable Common Agricultural Policy. Here discussion centres on the differences between EC and national-level policy making, and uses an amalgam of the several separate analytical models that have been developed elsewhere to describe the general decision-making process.[1]

The Policy-making process
Conventionally, the policy-making process is viewed as sequential or stepwise in character. In a national context, the process has its origins in the set of values which the society holds. Values can be thought of as comprising a complex of beliefs, preferences and aspirations that are held regarding what is desirable, the ultimate value perhaps residing in the well-being of each individual as a distinct and significant item of humanity.[2] Many societies have their underlying value systems formally expressed in a constitution, but usually values are translated into more tangible beliefs about how the economy or society should be organised. Values interpreted for the economic system might refer to the improvement of living standards or economic progress, balanced growth, employment opportunity, or economic stability and security. Values for the social system might concern distributive justice or an equitable remuneration for those employed in each sector of the economy. Equity values in agriculture, for example, have been widely

translated into income parity with the non-farm sector, together with income stability. Most societies comprising the EC hold similar values – indeed this feature has been identified already as one condition for a successful supra-national organisation – but they have been subject to redefinition through time and vary in relative importance between the different societies.[3] A major distinction, for example, can be drawn between the societies in which rural as compared with urban interests predominate. In the 1950s, France and Italy accorded a high priority to rural values in their policy making as befitted countries in which over twenty per cent of the working population earned their living from agriculture (Table 1). Farm people in particular were thought to make a special contribution to political, economic and social stability, economic growth, and social justice. In extreme cases, the ideals and virtues of an agrarian society are held as best for civilisation, the term 'rural fundamentalism' being used to describe such beliefs. Very often, rural fundamentalism is interpreted in terms of maintaining and strengthening the family-size farm, and the belief that the ownership of land in small parcels is the basis for a vigorous democracy retains its influence in many countries of the EC, particularly France. As urban rather than rural interests have pervaded each society, however, so the strength of rural fundamentalism has waned, and with the falling size of the rural population

Table 1. The importance of agriculture in Member States, 1958–82

Country	1958		1968		1973		1982	
	A	B	A	B	A	B	A	B
Greece							30.7	17.4
Ireland	38.4	26.0	29.4	18.8	24.8	18.5	17.1	11.1
Italy	34.9	18.5	22.9	9.9	18.7	8.6	12.2	6.3
France	23.7	10.2	15.7	7.5	11.1	7.1	8.2	4.3
Luxembourg	17.9	8.8	12.2	4.6	9.5	3.7	4.7	3.4
Denmark	15.9	16.0	12.8	7.7	9.5	6.7	8.4	5.5
West Germany	15.7	7.5	9.9	4.4	7.5	3.1	5.4	2.2
Netherlands	12.6	10.7	7.9	6.9	6.8	5.7	4.9	4.4
Belgium	9.4	7.4	5.6	4.9	3.8	4.1	2.9	2.6
United Kingdom	4.4	4.4	3.5	3.0	2.9	2.9	.7	2.3
EC (average of above)	19.2[b]	12.2[b]	12.0[b]	6.3[b]	8.3[c]	5.2[c]	7.6[d]	3.9[d]

A: Proportion of the working population employed in agriculture, forestry, and fisheries; B: Share of agriculture in GDP[a]
a: Gross domestic product formation at factor costs; b: The Six; c: The Nine; d: The Ten.
Source: Commission of the European Communities, *The agricultural situation in the Community, 1983 Report* (European Communities, Brussels, 1984) p. 187.

in the 1960s and 1970s there has been a parallel reduction in the priority accorded rural values throughout the Community.

The second stage of the policy-making process arises from the divergence between reality and the beliefs held by the society of how it and the economy should be organised. Typically, dissent about the existing state of the economy is expressed by sectional interests and it is translated either by pressure groups, political parties or the national legislature into policy objectives. Conveniently, the objectives of national agricultural policies tended to be very similar when the Community was formed. Thus while debate continued amongst the Member countries over specific policy measures, a short list of policy objectives for agriculture was incorporated into the Treaty of Rome reflecting the common interest. The objectives were broadly stated, to the frustration of those who subsequently have sought to determine if they have been reached or not. The advantages of broadly defined objectives, however, include the greater initial facility for political agreement, each country being at liberty to define the objectives more precisely for its own ends. In addition, objectives which are generally framed give greater freedom of action to individual members of society. The objectives co-ordinate their decisions and actions rather than direct their behaviour as happens in a command economy. Non-specific objectives also give greater freedom of action to governments when faced by changing social and economic conditions. New emphasis can be given to the different objectives but without recourse to altering the basic legislative framework.

The objectives of the CAP were stated as:

1 To incorporate agriculture within the common market (Article 38 (1)).

2 To increase agricultural productivity by promoting technical progress and by ensuring the national development of agricultural production and the optimum utilisation of all factors of production, particularly labour (Article 39 (1a)).

3 To ensure thereby a fair standard of living for the agricultural population, particularly by the increasing of the individual earnings of persons engaged in agriculture (Article 39 (1b)).

4 To stabilise (agricultural) markets (Article 39 (1c)).

5 To guarantee regular (food) supplies (Article 39 (1d)).

6 To ensure reasonable prices in (food) supplies to consumers (Article 39 (1e)).

7 To support the harmonious development of world trade (Article 110).

In addition, the objective of achieving a common organisation of agricultural markets was set out in Article 40 including the broad methods by which this was to be achieved. Non-specific 'price controls, subsidies . . ., arrangements for stock-piling and carry-forward, and common machinery for stabilizing importation or exportation . . .'[4] were all mentioned, together with a 'common price policy' and 'one or more agricultural orientation and guarantee funds'.

There is an inherent conflict between many of these objectives, for example between the statements on standard of living for farmers and food prices for consumers. 'Fair' and 'reasonable' in each case are open to a variety of interpretations, but both objectives cannot be met simultaneously, except at inordinate cost. Experience suggests that price levels which satisfy farmers are higher than those acceptable to consumers, and the converse, so that agricultural policy can end up satisfying neither group.

An environmental objective for agriculture is noticeable by its absence. Environmental issues were not of importance in the 1950s, and the general need for an environmental policy was formally recognised only at the 1972 'summit' conference in Paris. In November of the following year, the Council of Ministers adopted a number of objectives and general principles for Community activity which can be summarised as:[5]

1 To prevent, reduce and as far as possible eliminate pollution and nuisances.

2 To maintain a satisfying ecological balance and ensure the protection of the biosphere.

3 To ensure the sound management of and avoid any exploitation of resources or of nature which cause significant damage to the ecological balance.

4 To guide development in accordance with quality requirements, especially by improving working conditions and the settings of life.

5 To ensure that more account is taken of environmental aspects in town planning and land use.

6 To seek common solutions to environmental problems with states outside the Community, particularly international organisations.

The first three action programmes (1973–6 and 1977–82 and 1982–6) put into effect by the EC Environment Ministers have had their major impact on the urban and industrial sectors of the Community. Agriculture has been influenced only in a minor way by EC legislation, for example by limitations on the use of certain chemicals in pesticides and fungicides, since basic research work is incomplete on such issues as the environmental impact of intensive farming methods, the misuse of insecticides and certain fertilisers, intensive stock breeding, modern methods of cultivation, and the impact of land improvement schemes. Until this basic research work is completed, and proposals formulated from the results, national rather than Community regulations covering the rural environment will continue to have the greater influence.

The free market is incapable of providing a simultaneous achievement of the wide variety of policy objectives, and most democratically elected governments intervene in the market to regulate and control economic forces, but in a positive way so as to guarantee individual freedom. In effect, governments attempt to correct the structural defects of a market system. In this respect the EC has not acted differently from national governments throughout the world. Additionally, however, national governments have to take account of the different priorities accorded to the policy objectives by various groups within the society. Consumers, for example, tend to press for low-cost food from an efficient agricultural industry. But generally an efficient agricultural industry can be maintained only by transferring farm labour to other sectors of the economy. Conflict naturally arises between urban-orientated consumers and the farm population since the social and economic stability of rural areas is threatened. Groups vary also in their political power and influence so that compared with the well-organised and vociferous farming lobby, the consumer, until recently, has had few formal organisations through which to influence policy makers. A second reason for government intervention, therefore, is to ensure that in meeting the priorities of one group in the society other groups are not unduly disadvantaged.

This institutional view of government intervention is over-simplified for sectional claims are evaluated by national governments in relation to notions of a general 'national interest'. This somewhat nebulous concept often provides a national government with the justification for not meeting in full the demands of a particular group. In the EC, however, a comparable 'Community interest' remains weakly developed. Without this constraint national interests can be pursued at the expense of the wider Community, while the demands of sectional interests are more difficult to resist. Together these features exacerbate the problems of making sound agricultural policies in the EC.

The third stage of policy making attempts to resolve the conflicts between sectional interests and at the same time frame policy measures to reach the various policy objectives. National government ministers are commonly charged with the task of formulating legislative proposals which, after debate and correction by a democratically elected Parliament, eventually become law. But at this point the political process in EC policy making diverges markedly from practices within individual Member States and requires closer analysis.

The European Commission

The institutional structure of policy making for agriculture in the EC is shown in Figure 1, but in simplified form. The structure is similar for most areas of policy making but a diagram cannot represent adequately the very complex procedures of consultation and lobbying that in practice characterise the work of the Community. The traditional dialogue between Parliament and the Executive has been replaced by a dialogue between the two central institutions: the European Commission and the Council of Ministers. The Commission alone has powers to initiate and implement legislation, its strategic role reflecting the influence of 'federalists' on the institutional structure of the Community. Proposals are sent from the Commission to the Council of Ministers for decision. The Commission, however, is composed not of elected representatives but of fourteen semi-permanent officials and their staffs appointed for terms of office of four years by agreement between the Member States. Initially, the Commission acted as a collegiate body with a collective responsibility. But in recent years Commissioners have tended to operate as a collection of individuals with particular responsibilities, so wide has the range of Community activities become.[6] Two Commissioners are drawn from each of the four largest countries (France, Italy, West Germany, United Kingdom) and one from each of the others. Although Commissioners are supposed to act independently of any national interest, some observers have detected a growing nationalisation in the work of the Commission.[7]

Each Commissioner usually handles a specific aspect of Community business. Thus one Commissioner is responsible for agriculture, and the Directorate-General (Department) for Agriculture (D–G VI), with business subdivided into eight Administrative Units covering matters from international affairs relating to agriculture, through the organisation of markets for crop products, to agricultural legislation. Typically, an agricultural policy proposal is formulated within the Directorate-General for Agriculture. The 'leadership' provided by this body, and particularly by one of its early Agricultural Commissioners, Sicco Mansholt[8], has been an important factor in the progress achieved in developing a CAP[9]. However, the Commission's right of initiative is by no means as exclusive as often suggested. Externally stimulated policy proposals are also promoted by the EC federations of national professional associations and special interest groups, by public debate, and by complaints and requests from interested parties in the European Parliament and Member States. On occasion, the Council of Ministers proposes a 'resolution' which effectively instructs the Commission to act, although all proposals must pass through the Commission.

Initially, Commission staff consult government departments in Member States, agricultural experts, *ad hoc* working parties and pressure groups before completing their first study documents. Consultations include one of the important peripheral organisations of the EC – the Advisory Committees (Figure 1). For agricultural matters, the Committees are arranged on a product basis, but they also cover social conditions and farm structure[10]. Memberships of the various Committees are divided evenly between representatives of farmers, and those of food manufacturers, consumers, and trades unions. Their function is to keep the Commission aware of the practical implications of any proposal. These consultations serve to eliminate many areas of potential conflict although they also result in the curtailment of more radical, but possibly necessary proposals. Moreover, the extent of intergovernmental consultations, and those between the Commission and individual government departments, means that in practice a large proportion of the staff of national agricultural departments become the agents of the Community in developing and operating the CAP.

The preliminary draft of a policy proposal is then considered by the Agricultural Commissioner. When the draft has been approved matters may be dealt with through written documents or in discussion by a full meeting of the Commission. The draft can be further amended or accepted by the Commission, if necessary, by a simple majority vote, before being passed on to the European Parliament and the Council of Ministers for consideration. The Council undertakes its own series of consultations through working groups, and it is unusual for there not to be a conflict of interest between the Commission and the Council. The Commission, therefore, can adjust its proposals in the light of opinions subsequently expressed in the Council and thus act as a mediator between the Member States. As a conciliator of national viewpoints, the Commission removes contentious aspects from the policy proposals and there is an inbuilt tendency towards minimalist positions with each round of consultation and compromise.

The Commission fulfils two further roles in the EC once a decision has been taken by the Council of Ministers. First, the Commission takes the day to day management decisions in implementing and operating the CAP through the Directorate-General for Agriculture. In matters of administration or

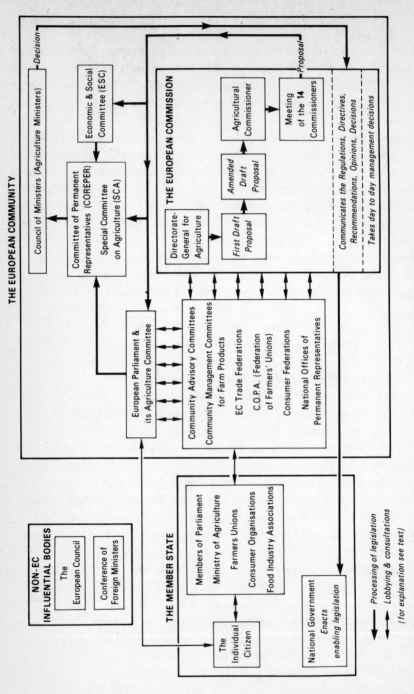

Figure 1. The institutional structure of policy making for agriculture in the EC

subordinate legislation the Commission is empowered to act independently of both the Council and the Parliament through its Management Committees. There is one Committee for each main group of farm products, each Committee consisting of civil servants drawn from Member States under the Chairmanship of a senior Commission official. The Committees advise and approve regulations and fix levies and other details associated with the CAP. Voting is by qualified majority, but if a Committee votes against a Commission proposal, the Commission can still take a contrary decision which remains in force unless the Council decides against it within one month. Secondly, the Commission is the 'guardian' of the Community Treaties, if necessary taking legal action through the EC Court of Justice against firms or governments infringing the legislation. These roles emphasise the pre-eminence vested in the Commission as a supra-national institution by the founders of the EC. In practice, however, the expression of national positions and the defence of national interests in the Council have remained central to decision taking on agricultural as on other important issues.

The Council of Ministers

Proposals from the Commission are passed to the Council of Ministers for executive decision. Foreign Ministers comprise the Council when it meets as a General Council, but it can convene also as a Technical Council on matters such as agriculture, industry, or energy. In this form, the Council is composed of government ministers responsible for the subject under review. National Ministers of Agriculture thus form the Agriculture Council. The Presidency of each Council, and its dependent bodies, rotates in alphabetical order amongst the Member States at six-monthly intervals. Before the Council considers a proposal, however, it is passed for discussion to a number of working groups and Committees from which the Council derives a part of its predominance over the Commission[11]. Each Member State can exert influence on Commission proposals through its national officials who comprise these committees, and significantly, their work has been dominated by agricultural matters.

Most policy proposals for agriculture are passed by the Council Secretariat to the Special Committee for Agriculture (SCA) which establishes their political feasibility, offers opinions and recommendations, and prepares the Agriculture Council's discussions. The SCA has operated since 1960 with over twenty sub-committees, and is composed of the agriculture heads of the permanent delegations to the EC together with senior officials of the national ministries and representatives of the Commission.

Certain agricultural matters – the harmonisation of legislation, financial issues, and commercial questions concerning third countries – and the wider business of Council sessions are prepared by the Committee of Permanent Representatives (*Comité des Représentants Permanents* – COREPER). This Committee is composed of the Ambassadors of Permanent Representatives of Member States to the Community and, while operating from 1958, was given official status by Article 4 of the 1965 'Merger Treaty'. The Committee incorporates a blend of national and EC loyalties thus acting both as a channel for expressing national reactions to Commission proposals, and as an interface between the Council of Ministers and the Commission in the consultations and negotiations that precede the adoption of any Commission proposal. The business of the Committee is subidivided between more technical matters dealt with by depty Permanent Representatives (COREPER I), and politically important issues reserved for the Ambassadors (COREPER II).

In practice, the SCA and the COREPER take many of the less controversial decisions on behalf of the Council, a feature of decision making that gives these Committees a key role in Community affairs. After debate, a draft proposal can go to Council as a 'point A' on the agenda when it can be adopted without debate. As a 'point B', however, a contentious proposal is subject to full debate and the Council can either act upon or ignore the advice of the SCA and the COREPER. Council meetings are the final forum of conciliation and compromise between national positions in which Commission officials also play their conciliating role. However, if no agreement can be reached the proposal may go back to the COREPER or the SCA, and the Council may have to give reasons to the European Parliament should it be disregarding its advice. In this circumstance, the Parliament may try to influence policy through further consultations with the Commission and relevant working groups of the Council.

Commission proposals can also be passed to the Economic and Social Committee (ESC) which is the official body through which economic and social groups express their views to the Commission and the Council. The ESC is comprised of 156 representatives drawn from socio-professional and consumer groups, employers' organisations, and trade unions in equal numbers. Twenty-four seats each are allotted to representatives from France, Italy, the United Kingdom, and West Germany, twelve each from Belgium, Greece and the Netherlands, nine each from Denmark and Ireland,

and six from Luxembourg. Appointments are made by the Council of Ministers, and on some matters consultation of the ESC is obligatory while on others it is optional. The ESC's agriculture section – one of nine such sections – has also taken initiatives in producing reports and opinions of its own in addition to discussing agricultural policy proposals and sending its opinion to the Commission and the Council through the COREPER or the SCA. The role and influence of the ESC, however, remain relatively limited.

Since 1974 the regular meetings of the heads of government, as the European Council, have acted as a further, and some would argue significant[12] influence on Council decision making. Although these 'summit' conferences cannot take legislative decisions, the European Council nevertheless passes agreements reached at the meetings to the Council of Ministers for ratification and implementation. One example of this process is provided by the agreement of the Dublin 'summit' of March 1975 to continue to import certain quantities of New Zealand butter, a decision subsequently implemented by the Agriculture Council in June 1976[13]. In a similar way, the Conference of Foreign Ministers meet regularly to co-ordinate the foreign policies of the Community.

After debate, the Council can act by a 'qualified majority' vote and has done so on budgetary matters and in the Agriculture Council when a compromise solution to a problem has not been forthcoming. A qualified majority is formed by 45 votes or more from a total of 63, but if the proposition being voted on emanates from elsewhere than the Commission the votes must be cast by at least six countries. The votes, weighted according to population size, are ten each from France, Italy, the United Kingdom and West Germany, five each from Belgium, Greece and the Netherlands, three each from Denmark and Ireland, and two from Luxembourg. Denmark and Ireland are particularly advantaged under qualified majority voting and there is a general weighting in favour of small countries[14]. Opposition to the supra-national implications of majority voting was one issue which led to the withdrawal of France from the EC between July 1965 and January 1966. The other issues concerned Commission proposals on the method of funding the Community's Budget, and on extensions to the supervisory powers of the European Parliament with regard to the Commission's activities. The 'Luxembourg Compromise', which ended the 'empty chair' crisis, informally reserves powers of veto to each Member State when a special national interest is claimed. In general, the Council has avoided majority voting[15], while in any case unanimity is required to amend a proposal from the Commission. Coalitions offer few benefits to any country and the threat, if not the practice of veto when sparingly used has been the most potent instrument for furthering an individual national interest. In 1963/4, for example, France threatened to withdraw from the EC unless West Germany agreed to harmonise grain prices. The threat was sufficient to produce an agreement in December 1964, although arrangements were not concluded until July 1967.

Undoubtedly, domestic politics influence negotiating stances and tactics since domestic political advantage can be gained by appearing to defend a national interest in Council sessions. An example of this feature was provided in 1971 by the objection of Italy to proposals for granting preferential trade conditions to Morocco and Tunisia. Members of the Italian Parliament and government were under constant pressure from the rural constituencies in the south and in Sicily to protect their fruit and vegetable production. In the event, Italian objections held up negotiations but could not prevent the Maghreb Association Agreement being concluded. A second example is provided by West Germany where, for many years, the government was formed by a coalition of the Social Democratic Party (SPD), led by Chancellor Helmut Schmidt, and the Free Democratic Party (FDP). The FDP, and hence the coalition, was dependent upon votes from the rural areas. Consequently, Joseph Ertl of the FDP, as a long-serving Agriculture Minister, had a domestic political interest in defending those aspects of the CAP which were favourable to farmers in West Germany.

In practice, the definition of what constitutes a national interest has varied from country to country and through time, often depending on the pressure on the governing political party for electoral survival. In addition, national governments have not attempted to maximise their interests separately in each policy sector. A Member State may sacrifice some of its interests in one field if it is compensated in other areas. The exchange of costs and benefits in defining interests in this way are calculated in both political and economic terms.

Four types of EC legislation can be drawn up following approval of a proposal by the Council of Ministers. A 'Regulation' is a Community law which supercedes and overrides national law. Regulations are applicable throughout the Community and do not require further national legislation. EC agricultural prices, for example, are made effective by Regulations. A 'Directive' is binding in the aim to be achieved but is applied by each national government in such a way that it fits into existing national legislation. Directives are generally issued when the different consititutional structures of the Member States makes it impossible or impracticable for a

Regulation to be issued. 'Decisions' are usually of a more limited or specific practical character often relating to a single Member State but binding upon it. After Regulations, Decisions are the most important category of rule-making in the EC. 'Recommendations' and 'Opinions' have no binding force, nor do 'Resolutions', but they may be significant in laying down principles for future action. When a text has been agreed by Council it is sent to the *Groupe des Juristes-Lingues* for translation into the official languages of the EC and then published in the Official Journal if it is a Directive or Regulation.

Conclusion

Agricultural policy reflects the values of the society for which it is made but it is also conditioned by the nature of the policy-making process. In the EC, the process departs from practices in individual Member States in a number of important ways which in general heighten the problems of making sound agricultural policies. For example, EC policy making is far more complex than at the national level, involving a considerable degree of consultation, negotiation and compromise between the national bureaucracies. In some respects, therefore, policy making in the EC can be interpreted as a close representation of the inter-governmental model of international relations where formal contacts between governments are made through a central authority.[16] However, detailed retrospective studies of specific policy decisions taken by the Council have revealed the operation of a variety of processes.[17] For some agricultural policies, the role of the Directorate-General for Agriculture and the Commission has been of critical importance, and the freedom of action of governments in practice has been limited by a complex sharing of power between national and Community institutions. So complex has the consultation process become that apparently inordinate amounts of time are spent on committee work. Moreover, since many decisions are taken in committees, national civil servants have come to play an enhanced role in Community as compared with national policy making, to the extent that some decisions owe more to the influence of élite networks – small groups of closely integrated officials – than to the inter-governmental bargaining. Thus, power and authority in the EC are not fixed but shift amongst the policy-making bodies with the issue under debate.

Member governments nevertheless retain a primacy in Community policy-making. In the present context they shape the demands of the agricultural industry at the national level before transmitting them to the Community through the Council of Ministers and its dependent committees, especially the SCA.[18] This structure allows narrowly defined national interests to be pursued, and the search for unanimity in inter-governmental bargaining leads to short-term, compromise decisions not conducive to the making of sound agricultural policy. Most governments tend to treat each issue on its individual merits, hesitating to accept agricultural policies which may either have far-reaching effects on national agricultural systems or lead to pressure for change in other areas of Community activity. Long-term agricultural objectives tend to be overlooked or avoided, while periodic reorientations of agricultural policy, so necessary in the face of changing economic and political conditions, are extremely difficult to achieve. Inertia in the CAP, therefore, has tended to be greater than in the national agricultural policies which it replaced, with an inevitable drift towards incrementalism and continuity.

Notes

1 G. C. Rosenthal, *The men behind the decisions* (Massachusetts, Lexington, 1975), pp. 3–7.
2 W. R. Parks, 'Goals of democracy', in R. J. Hildreth (ed.), *Readings in agricultural policy* (Lincoln, University of Nebraska Press, 1968), pp. 3–12.
3 This theme is developed for the UK in I. R. Bowler, *Government and agriculture: a spatial perspective* (London, Longman, 1979), pp. 41–54.
4 Quoted by R. Fennell, *The Common Agricultural Policy of the European Community*, (London, Granada, 1979), p. 9.
5 Commission of the European Communities, *The European Community's environmental policy*, European Documentation 1977/6, (Brussels, European Communities, 1977).
6 H. Wallace *et al* (eds.), *Policy making in the European Communities* (London, John Wiley, 1977), p. 311.
7 W. F. Averyt, *Agrapolitics in the European Community: interest groups and the CAP* (New York, Praeger, 1977), pp. 84–5.
8 The Agriculture Commissioners to date have been in sequence: Sicco Mansholt, Petrus Lardinois, Finn Gundelach, Poul Dalsager.
9 L. Lindberg and S. Scheingold, *Europe's would-be polity* (New Jersey, Prentice Hall, 1970), pp. 141–81.
10 Economic and Social Committee of the European Communities, *Community Advisory Committees for the representation of socio-economic interests* (Farnborough, Saxon House, 1980), pp. 49–51.
11 H. Wallace, *National governments and the European Communities*, PEP (London, Chatham House, 1973).
12 J. Cooney, *A United State of Europe?* (Dublin, Dublin University Press, 1980), p. 19.

13 M. Tracy, 'The decision-making process in the European Community with reference to agricultural policy', in T. Dams and K. E. Hunt (eds.), *Decision making and agriculture* (Lincoln, University of Nebraska Press, 1977), pp. 323–40.

14 R. J. Johnston and A. J. Hunt, 'Voting power in the EEC's Council of Ministers: an essay on method in political geography', *Geoforum* VIII (1977), pp. 1–10.

15 This practice was dramatically suspended on May 18th 1982 when the UK's veto on proposed price increases for agricultural products was overturned. In effect the price proposals were accepted by a qualified majority vote.

16 R. Dahrendorf, 'A new goal for Europe', in M. Hodges, *European integration: selected readings* (Harmondsworth, Penguin, 1972), pp. 74–87.

17 Rosenthal, *The men behind the decisions*, p. 80.

18 Fennell, *The Common Agricultural Policy*, pp. 50–1.

3
Policy making for Agriculture: the agricultural interest

Within the EC, the agricultural sector enjoys a political influence out of proportion to the employment or GDP which it generates. To an extent this is a reflection of the political leverage that agriculture can exert within individual Member States. In West Germany, for example, leverage is exerted by the pivotal political position of the rural vote, while in Denmark it stems from agriculture's role in the economy as an export earner. For a variety of domestic political, social and economic reasons, therefore, many national governments press the case of agriculture in the Council of Ministers to an extent not enjoyed by other sectors of the economy.

In addition, the agricultural sector has been extremely active at the EC level in exploiting the network of consultations that are a central part of policy making. Here the political influence of agriculture is traced through the work of the European Parliament and EC-level pressure groups.

The European Parliament
The Council, in its policy making, is not subject to direct control by a democratically elected body and in this respect the EC departs from practices in most Member States. Nor is the national bargaining inherent in Community policy making beneficial to the democratic process within Member States. Each Minister is undoubtedly accountable to his or her own national government and ultimately to the national parliament and electorate. But in national parliamentary debate and questioning before Council negotiations ministers are reluctant to disclose their bargaining hand, while afterwards it is too late for parliaments to influence the decisions.[1] Most decisions, once taken, are binding on all Member States which have no power to modify them. They can be reversed with only the greatest difficulty since in general any proposal before the Council must

emanate from the Commission. This body has to be persuaded of the need to alter a decision before action is possible. Neither is the Commission subject to direct democratic control, in the normal sense of the term, and certainly not in the way that a civil service, its closest parallel, is subject to control in national arrangements.

The European Parliament, sometimes described as the 'federalist heritage',[2] might be expected to occupy the role in EC policy making that national parliaments, or their equivalent, fulfil in the affairs of Member States. The powers of the Parliament, however, are closely circumscribed owing to their supra-national implications. Significantly, the Parliament was termed a 'Common Assembly' in the early years of the Community and the direct election of its 410 members (MEPs) took place for the first time only in June 1979.[3] Previously, its 198 seats had been filled by delegates from Member countries' own parliaments, and were judged by many merely to represent particular interest groups. For example, in April 1978, in a debate on proposed price increases for agricultural products, those MEPs representing producers voted for the price increases while those representing consumers – urban MEPs – voted for a freeze on prices: the divide crossed political groups and Member States.[4]

The Council refers most legislative and all budgetary proposals of the Commission to the Parliament for scrutiny and opinion (Figure 2). Although the Parliament has the right to be consulted on most matters, it has no powers to initiate or amend legislation. Rather it can only urge the Council to adopt a proposal, or influence the Commission to alter its proposals. The Commission nevertheless retains the right to reject any recommendation from the Parliament. In effect, the Parliament provides a forum in which proposed legislation, and the activities of the Council and the Commission, can be subjected to observation, questioning – both written and oral – and objection. Both the Commission and Council members take part in parliamentary debates and 'question times' in which they are open to cross-questioning and hostile criticism. Unlike the Commissioners, however, representatives of the Council are not obliged to answer questions, although the Council, in reaching its decisions, has shown itself to be responsive to the opinions offered by the Parliament. Nevertheless, the control powers of the Parliament are limited to being able to dismiss the Commission as a whole on a motion of no confidence, and is the only body to do so. However, a two-thirds majority of those voting is required together with an absolute majority of the Parliament's members. The Parliament can also reject proposed expenditure under the Community's Budget, but these remain blunt instruments of control. For example, individual Commissioners cannot be dismissed, while only four votes of censure have been moved in recent years. Two votes were taken – in June 1976 and March 1977 – and in both cases the motion of censure was defeated.

When the President of the Parliament receives a Commission proposal, it is allocated to one of fifteen standing committees.[5] The Agriculture Committee, generally comprised of members with some experience of the farming industry, prepares a draft report on the proposal through its rapporteur. The draft is then debated, amended and voted upon in a full session of Parliament with Commissioners present. The opinions and recommendations of the Parliament are then returned to the Commission and the Council (Figure 2). Some policy proposals, however, may be accepted without a report (5 per cent of the total) and some draft reports may be accepted without debate (9 per cent of the total).[6] The Commission, in the light of Parliament's Opinion, may send a revised proposal to the Council, and it reports back to the Parliament on the action it has taken. In the past, Commission proposals and Parliamentary resolutions have tended to support Community action, and it has been the Council that has shown reluctance in taking the necessary decisions. For example, in the year November 1977 to October 1978 Parliament delivered 160 Opinions on Commission proposals. In 7 per cent of cases Parliament adopted the proposal without amendment, in 19 per cent of cases the Commission accepted proposed amendments, and in only 9 per cent of cases were proposed amendments not accepted by the Commission. These figures, of course, say nothing about the relative importance of matters on which Opinions were delivered. When the Parliament delivers a negative opinion on a proposal, especially one which would require funds to be voted on by Parliament, the Council and the Commission reconsult the Parliament and attempt to arrive an an acceptable compromise (Figure 2).

The agricultural or rural interest is represented in Parliamentary activity through the national political parties. MEPs, however, sit in Parliament according to six main political groups rather than national delegations, and to a large extent the unequal allocation of seats among the Member countries, as judged by the ratio of population per seat,[7] is subservient to the uneven representation of the different political groups (Table 2). At one level, a national political party can exert a large influence over a particular group, for example, the Italian Communists over the Communist group; but at a second level, the groups vary in power within the Parliament. The Socialists have the largest number of MEPs followed by the European People's Party

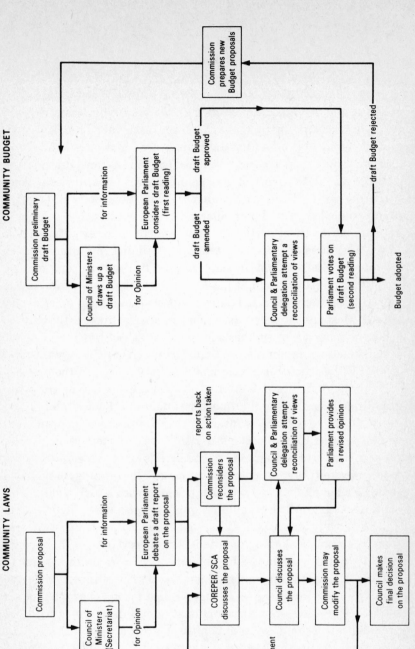

COMMUNITY LAWS

COMMUNITY BUDGET

Commission proposal

for information

Council of Ministers (Secretariat)

for Opinion

European Parliament debates a draft report on the proposal

reports back on action taken

Commission reconsiders the proposal

Council & Parliamentary delegation attempt reconciliation of views

Parliament provides a revised opinion

COREPER/SCA discusses the proposal

Council discusses the proposal

Commission may modify the proposal

Council makes final decision on the proposal

no agreement

Commission preliminary draft Budget

for information

Council of Ministers draws up a draft Budget

for Opinion

European Parliament considers draft Budget (first reading)

draft Budget approved

draft Budget amended

Council & Parliamentary delegation attempt a reconciliation of views

Parliament votes on draft Budget (second reading)

Commission prepares new Budget proposals

draft Budget rejected

Budget adopted

Source : Commission of the European Communities

Figure 2. The role of the European Parliament in decision-making

(EPP); but a formal alliance has been forged between the European Democrats, the Liberals and Democrats, and the EPP to create a Parliamentary majority grouping. However, with the enlargement of the Community in 1973, and the direct election of 1979, a stimulus has been given to the creation of federations amongst these transnational political groups.[8]

Most of the groups have internal divisions based on ideology or national interests. Thus while they have all developed common policy positions, the groups vary greatly in the depth of their commitments to particular policies and in the degree of detail with which policies have been elaborated. Agricultural policy has proved particularly divisive. Most groups contain at least one national political party that is dependent to an extent upon the rural vote, and tensions are created by a conflict of interest with urban-orientated MEPs within the same group. The resulting compromises have meant that no party grouping has been able to elaborate a farm policy which differs essentially from the agreements reached by Member States over the CAP. Most groups, therefore, either lend their support to the CAP, or wish to see only changes in emphasis in the existing policies.

The Liberal group, for example, obtaining many of its MEPs from the rural areas of the EC, has the strongest agricultural interest and in the past has been active in the Agriculture Committee of the Parliament. The group has had a firm attachment to small farmers, has doubted the desirability and practicality of the structural solution to the 'farm problem', and has emphasised the need to take account of rising costs of production in price setting. The inclusion of British and Dutch Liberals in recent years, however, has tended to weaken the group's unified support for the farming community. The new MEPs have tended to favour a pricing policy which is of greater benefit to the consumer.[9] The Communist group has also spoken in defence of the small farmer but for ideological reasons. The CAP, with its structural component, is seen to favour large capitalist farming systems at the expense of the small producer. The group favours a wide-ranging system of co-operatives which would leave ownership patterns intact but give flexibility in meeting technological change without recourse to high product price levels. Nevertheless, the Communist group has no agricultural programme but rather pursues 'common themes' for action.

The European Progressive Democrat group (EPD), a coalition of French Gaullist and Irish Fianna Fáil MEPs, has also supported the CAP. Their 'Agricultural Charter' of 1975 emphasised the need to consolidate rather than revise the CAP, and called for stronger market organisation together with a structural policy aimed at maintaining a balanced agricultural population.

Table 2. Political groups in the European Parliament (1979 elections)

Country	% of electorate voting[a]		Total seats	Population per seat ('000)	Political affiliation (seats in each group)						
					S	EPP[b]	ED[c]	C	L	EPD	Others[d]
West Germany	65.7	(90.7)	81	765.9	35	42	0	0	4	0	0
United Kingdom	32.6	(76.0)	81	692.0	18	0	61	0	0	1	1
Italy	85.5	(89.9)	81	683.5	13	30	0	24	5	0	9
France	60.7	(82.8)	81	664.0	22	8	0	19	17	15	0
Netherlands	57.8	(87.5)	25	538.0	9	10	0	0	4	0	2
Belgium	91.4	(94.6)	24	407.2	7	10	0	0	4	0	3
Denmark	47.8	(88.9)	16	315.8	4	0	3	1	3	1	4
Ireland	63.6	(76.9)	15	205.7	4	4	0	0	1	5	1
Luxembourg	88.9	(90.1)	6	59.5	1	3	0	0	2	0	0
EC 1979	72.8	—	410	631.6	113	107	64	44	40	22	20
EC 1978	—	—	198	1307.9	66	53	18	18	23	16	4

C: Communist and allies; S: Socialist; EPP: European People's Party; EPD: European Progressive Democrats; ED: European Democratic; L: Liberal and Democratic

a: figures in brackets are for most recent national elections; b: before 1976, the Christian Democrat Group; c: formerly the European Conservative Group; d: Independent and Group for the Technical Co-ordination and Defence of Groups and Non-attached members

Source: Directorate-General for Information and Public Relations

Traditionally, the large rural vote for the Christian Democrat parties (now the EPP) has led the group also to maintain a close interest in the CAP. Existing policies have been supported in part because the CAP is the only common policy in the Community, but more recently because of the group's support for the rather vague concept of 'modern family farms'. In the past, some powerful spokesmen for the agricultural interest have sat in the Christian Democrat group.[10] They have included, for example, the presidents of the *Association de L'Industrie Laitière de la Communauté Européene* (ASSILEC) – an organisation representing the milk processing industry, the *Comité des Organisations Professionelles Agricoles* (COPA) – a federation of national farmers' unions, and the Dutch Catholic agricultural workers' syndicate. From time to time, however, the group has been divided by regional and product interests. Italian members of the group, for example, have defended products such as wine, fruit, and tobacco in the CAP against the livestock orientation of German group members. Dutch members have argued against high farm prices and have voted with the Socialist group on such matters. Even national coalitions of groups have been formed over certain issues – the Dutch over cereal prices in 1962, for example, and the Italians over tobacco policy proposals in 1969.

The consumer interest, by contrast, has been taken up more strongly in recent years by the European Democratic group whose position in the Parliament was greatly strengthened after the 1979 direct elections (Table 2). Although the basic principles and existing mechanisms of the CAP are again supported, the group has attacked proposals for sizeable increases in farm prices. The Socialist group has gone so far as to elaborate a general political programme at the European level, issuing, as early as April 1961, an 'agrarian programme'. However, no farm programme was issued during the 1979 Parliamentary election campaign, and the group has been less sympathetic towards the agricultural interest than others. It has opposed excessive protectionism at the expense of the urban working class (consumers), and favoured structural rather than price policies for serving agriculture's long term problems. Nevertheless, tensions exist within the group: between the more urban-orientated members of the British Labour Party and West Germany's SPD, and those MEPs drawn from the French Socialist and Irish Labour parties who seek support for their farmers.[11]

Traditionally, therefore, the European Parliament has been sympathetic towards agriculture and has tended to support the CAP which, in its present form, favours the farming community at the expense of the consumer. There are, nevertheless, several trends which are reducing the traditional influence of the agricultural interest in the Parliament. For example, the falling number of workers in the agricultural industry has eroded the peasant base of many national political parties, and the increasingly urban character of the electorate has become a feature to be reckoned with both by the political parties and individual MEPs. Whether or not rural interests are proportionally over-represented in the Parliament by the way constituency boundaries are drawn has yet to be determined, but the large size of each constituency in any event reduces the influence of any single pressure group. Nor is the agricultural interest so strongly represented in the newly elected Parliament. Fewer than ten per cent of its members can be identified as having a direct interest or background in agricultural matters. Thus initially the elected Parliament showed every sign of being more critical of legislative and budgetary proposals than its predecessors, especially with respect to expenditure on agriculture. In autumn of 1979, for example, the Parliament refused to sanction increased agricultural expenditure, and ultimately rejected the whole of the proposed Budget for 1980 by a substantial majority (288 votes to 64 with one abstention). Subsequently, however, the Parliament proved less critical of price proposals under the CAP.[12]

Pressure groups

The Community system of policy making offers great scope to pressure groups to influence decisions. That this opportunity has been seized upon and developed reflects in part the need for national sectional groups to defend and represent their interests in a new centre of decision making, and in part the emerging consciousness among the electorate of the need for greater participation in decision making. Sometimes, the formation of EC-level pressure groups has been spontaneous, but on other occasions it has been prompted by an invitation or even by some pressure from the Commission.

Over 150 pressure groups have been identified which are concerned to some degree with the agricultural sector.[13] They vary between those that have some national advantage to gain on the basis of a common objective interest (sectional interest group) and those which are held together only by common ideals or policies (promotional interest or attitude group). A more useful distinction can be drawn between broad-sector groups such as the National Farmers Union (NFU) of the United Kingdom, the *Centre National des Jeunes Agriculteurs* (CNJA) of France, and the German Farmers' Union *Deutsche Bauernverband – DBV*) and narrowly focussed groups such as the National Federation of Milk Producers – *Fédération Nationale des Producteurs de Lait* of France. Some organisations, such as the NFU, have been recognised at the national level as the main consultative bodies representing the whole farming interest. Others have a more direct political attachment. In Italy, for example, the *Coltivatori Diretti* is a Christian Democrat organisation for farmers while *Confagricultura*, representing large landowners, is attached to the Liberal Party. The DBV exercises a direct influence on domestic agricultural politics. In the past the DBV was closely associated with the Christian Democratic parties. But with the political decline of the Christian Democrats after 1969, the DBV became more involved with the FDP and its alliance with the SPD. The Union has been able to mobilise the farm vote in pivotal rural – Protestant areas, and has had its own FDP candidates in national elections.[14]

Without doubt a large part of the influence of the agricultural interest in the EC is derived from these national farm groups. They lobby individual Ministers of Agriculture and their civil servants thus helping to establish national positions on policy issues. These positions are then represented at the Community level in decision-taking meetings of the Agriculture Council, the SCA, and the various Management Committees.

Another channel of influence goes directly to the Community authorities where the national groups can exert pressure individually. The Council, however, consults pressure groups only infrequently and then on an informal basis; the Commission for its part generally restricts its consultations at the discussion stage of policy formulation to officially recognised EC-level pressure groups (Eurogroups). So as to meet this condition, agricultural pressure groups have had to form EC-level organisations with interests that vary from food processing *Commission des Industries Agricoles et Alimentaires* – CIAA), through agricultural co-operation *Comité Général de la Co-opération Agricole* – COGECA), to farm production (COPA). The Eurogroups vary in size and influence. For example, the European Committee for Agricultural Progress, (*Comité Européene pour le Progrès Agricole* – COMEPRA) represents a number of small farm groups such as the National Alliance of Italian Farmers, the Action Committee of Walloon Peasants, the Democratic Farmers' Action of Germany, and the National League of Family Farmers of Ireland. COMEPRA, in contrast to COPA, opposes many aspects of the CAP. More specialised Eurogroups are exemplified by the Confederation of European Sugar Beet Producers (CIBE) which only represents a specific farm sector.

The Eurogroups on the one hand foster and promote the exchange of information so as to instil a spirit of co-operation and cohesion into their affiliates. On the other hand, they co-ordinate and exert pressure for policies through the central organisation and the national affiliates at both EC and national levels. A high proportion of the pressure groups maintain offices and officials in Brussels so as to keep in touch more easily with the activities of the Community, and to be on hand when lobbying decision makers. Certain groups have gained added influence through becoming the bodies officially recognised by the Commission for negotiation and consultation. COPA and COGECA fulfil this role for the agricultural sector, for example, and of the two COPA is widely considered to be amongst the most influential of all pressure groups in the Community.

COPA is a federation of over twenty national farmers' unions and its strength has four bases. First, national affiliates channel all their demands through the central organisation, and by agreeing a common position, a united front can be presented to the Community institutions. COPA is controlled by an Assembly composed of representatives from the member organisations, while a 'Praesidium' consisting of the presidents of the national unions, takes all decisions within the framework of the Assembly's general guidelines. Since 1973 COPA has made its internal decisions by qualified majority voting. This procedure has helped create both compromise positions, thus resolving internally national or sectoral divergences of interest, and speedy decisions which enhances COPA's ability to influence events. Such a procedure is not followed by all interest groups however. Some proceed strictly by unanimity, others by simple majority voting but allowing minority views to be expressed, and yet others by a simple majority without allowing minority views. Secondly, in common with most pressure groups, COPA has a secretariat. But in addition there is a well-developed structure of working groups comprised of officials drawn from the staffs of the member organisations. A 'general experts group' carries out studies of an economic or policy nature and co-ordinates the positions and recommendations of the working groups.[15] The working groups, themselves subdivided mainly on a commodity basis but including farm accounts and social matters, examine current policy in their respective sectors, and make recommendations to the praesidium. In addition the groups publish information reports and pamphlets thus maintaining a steady flow of communication between the central EC organisation and its affiliates. Thirdly, COPA (and COGECA) provide both the personnel for the Advisory Agricultural Committees, usually drawn from the officials in the relevant working groups, and representatives for the ESC.[16] In addition, the praesidium has regular meetings with the Commissioner for Agriculture and the Agriculture Committee of the European Parliament, while a network of informal contacts exists between members of the working groups and decision makers in the middle level of communication and decision in the EC institutions. Fourthly, the member affiliates are prepared to implement central decisions at the national level and so present individual Agriculture Ministers with a relatively uniform agricultural viewpoint.

A number of different 'strategies of influence' have been developed by national farm groups within the EC.[17] The orthodox strategy (Figure 3) involves lobbying the Commission through the channels developed by COPA, and reaching national governments through the contacts made by individual national farmers' unions. In an emergency situation, as for example when pressing for interim price increases in September 1974, COPA co-ordinates the lobbying of national governments. A national farm group can also act independently by placing pressure on a national government to solve a particular national farm grievance through the Council of Ministers. Alternatively, a national farm group can lobby its own government to introduce a national aid scheme. In 1974, for example, 165 such schemes were submitted by national governments to the Commission for approval.

Whether acting through the Council or the Commission, the agricultural interest for many years has exerted a considerable effect on the development of the CAP and on the annual farm price negotiations. COPA, for example, has been attributed with redirecting the Commission's important policy proposals on agriculture between 1969 and 1971 (the Mansholt Plan). The farm structure aspects of the proposals were de-emphasised in favour of attaching greater importance to price measures. COPA has also promoted policy initiatives as well as responding to Commission proposals. The recently introduced market organisation for sheep products, for example, stems in part from pressure by the agricultural interest. Pressure by farmers, however, has not always been confined to formal channels. In the past there have been many violent demonstrations at national levels against the CAP, and occasionally on the streets of Brussels itself, as in March 1971.

Consumers and environmentalists have also formed promotional interest groups at the EC level. In general, promotional groups are half as numerous as special interest groups and have been created mainly in the last decade. As yet they are not sufficiently well-organised or funded to counteract the influence of a group such as COPA, nor are they as fully represented in the EC institutions. The Treaty of Rome, for example, makes little mention of the consumer interest, for the 'consumer movement' in general did not gather momentum in Western Europe until the late 1960s. Consumers thus do not have a place as of right on the ESC, while places on the Advisory

The orthodox strategy

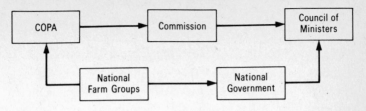

The emergency strategy

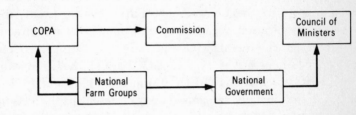

EC solution to a national grievance

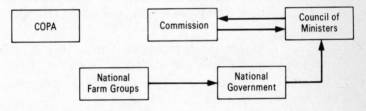

Retreat to national measures

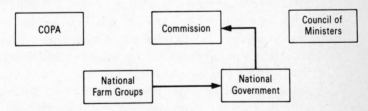

Source: Averyt, 1977

Figure 3. Strategies of influence by national farm groups within the EC

Agricultural Committees have to be shared with food manufacturers and trade unions.[18]

In January 1973, however, the Commission's Environment and Consumer Protection Service replaced the administrative unit for consumer affairs which previously had operated from the Commission's Competition Department. Its task is to look after the interests of consumers in the EC and it falls directly under the responsibility of a Commissioner. In September of the same year a Consumers' Consultative Committee (CCC) was set up which reports to the Commission. Fifteen of its twenty-five members represent organisations regarded by the Commission as representative of the consumer. Such organisations include the European Community of Consumer Co-operatives (EURO-COOP) and the European Bureau of Consumers' Associations (BEUC) formed in 1957 and 1973 respectively. BEUC, as the officially-recognised body representing the consumer interest, has developed informal contacts with the Commissioner for Agriculture, especially during the consultations leading up to the annual setting of agricultural prices. Four members of the CCC are experts appointed by the consumer organisations, and six are independent experts appointed by the Commission.

The pressure groups naturally come into conflict. BEUC, for example, has tended to campaign on behalf of consumers to severely limit increases in prices on those agricultural products in surplus production. COPA, by contrast, has presented the case for price increases consistent with the maintenance of agricultural incomes. In practice, neither group has been satisfied with the outcome: price increases have been higher than canvassed by BEUC and lower than desired by COPA.

However, pressure groups that are formally incorporated into the EC policy-making process can face a dilemma. A conflict can arise between a duty to their members in representing their interests, and their Community role as a mechanism whereby compromise is reached among the various national positions on a given policy proposal. Moreover, their effectiveness can be limited because the harmonisation and co-ordination of common interests among the member organisations is impeded by the heterogeneity of the economic and political systems they represent. Some observers, therefore, conclude that interest groups are greatly restricted in their activities and effectiveness at the EC level, but the agricultural interest should be excluded from such a generalisation.

Conclusion

This chapter has sought an explanation for the disproportionately large political influence wielded by the agricultural sector within the EC. Attention has been focussed on the network of lobbying and consultations formed by the policy-making process which allows close contacts to be developed amongst government ministers, Commission officials, and leaders of pressure groups. In this corporatist view,[19] the agricultural interest ensures that its opinions are represented to influential committees both at national and EC levels, that it has spokesmen in the European Parliament and the political groups, and that the individual Agriculture Ministers are well-briefed on its views. The Ministers carry their responsibility into the Council and in effect represent their national agricultural industries. Indeed, a considerable degree of autonomy has had to be granted to Agriculture Ministers, and from time to time they become relatively immune from interference from their national administrations, particularly during the annual negotiations on agricultural prices. Together these features have tended to magnify rather than diminish the problems of making sound agricultural policies. On the one hand, the checks and balances of the policy-making process tend to stifle necessary redirections of agricultural policy; on the other hand the agricultural interest has an enhanced ability to influence the development of policy in its favour compared with the national level of policy making.

In the European Parliament the agricultural interest is pursued through oral and written Parliamentary questions, but more particularly through the work of the specialised Agriculture Committee which deals with the pre-decision stage of the legislative process. With the growing urbanisation of the societies comprising the EC, however, urban rather than rural interests are becoming more influential in the development of agricultural policy. Several of the recent conflicts in agricultural price negotiations can be attributed to this shift of influence, with entrenched farming interests often having to defend the CAP against rising consumer criticism and the concern of Finance Ministers (since the 1980/1 agricultural prices settlement) to limit the costs of agricultural support. The Parliament, however, will need to develop a stable majority amongst its constituent party groupings if it is to deal with the problems of agricultural policy in the medium term. At present groups are still too prone to divide and reform according to ideological, political and national imperatives.[20]

Notes

1 Open University, *The European Economic Community – economics and agriculture* (Milton Keynes, Open University Press, 1974), p. 32.

2 P. Scalingi, *The European Parliament – the three decade search for a united Europe*, European Studies 3 (London, Aldwych Press, 1980), pp. 3–12.

3 The EC was enlarged from six to nine members in January 1973 by the inclusion of Denmark, Ireland and the United Kingdom. Membership further increased to ten countries in January 1981 when Greece joined the Nine, at which time the number of seats in the European Parliament was raised to 434. The additional twenty-four MEPs were initially appointed by the Greek national Parliament until elections could be held.

4 G. and P. Pridham, *Transnational party co-operation and European integration* (London, George, Allen and Unwin, 1981), p. 90.

5 The fifteen Committees of the European Parliament are: Agriculture (39), Political Affairs (41), Legal Affairs (25), Economic and Monetary Affairs (37), Budgets (37), Budgetary Control (27), Social Affairs and Employment (27), Youth Affairs/Culture/Education/Information and Sport (27), Regional Policy and Planning (29), Transport (25), Environment/Public Health and Consumer Protection (27), Energy and Research (34), External Economic Relations (36), Development and Co-operation (27), Rules of Procedure and Petitions (27). The figures in brackets indicate the number of MEPs on each Committee in 1979/80.

6 J. Fitzmaurice, *The party groups in the European Parliament* (Farnborough, Saxon House, 1975), p. 55.

7 R. J. Johnston, 'National power in the European Parliament as mediated by the party system', *Environment and Planning* IX (1977), pp. 1055–66.

8 Pridham, *Transnational party co-operation*, pp. 1–37.

9 S. Henig (ed.), *Political parties in the European Community* (London, George, Allen and Unwin, 1979), p. 260.

10 Fitzmaurice, *The party groups in the European Parliament*, p. 83.

11 Henig, *Political parties in the European Communities* p.253.

12 For example, in March 1983 the Parliament voted (by 147 to 123) against Commission proposals to control surpluses and in favour of a relatively high price increase for farm products: seven per cent as against the Commission proposal of 5.5 per cent. The vote reflected a growing opposition to the way in which the CAP is biased towards large at the expense of small farms.

13 R. Fennell, *The Common Agricultural Policy of the European Community* (London, Granada, 1979), p. 63.

14 E. Andrlik, 'The farmers and the State: agricultural interests in West German politics', *West European Politics* I (1981), pp. 104–19.

15 Fennell, *The Common Agricultural Policy*, p. 63.

16 Economic and Social Committee of the European Communities, *European*

interest groups and their relations to the Economic and Social Committee (Farnborough, Gower, 1980), p. 204.

17 W. F. Averyt, *Agrapolitics in the European Community: interest groups and the CAP*, (New York, Praeger, 1977), pp. 100–12.

18 European Communities, *The consumer in the European Community* (Brussels, Commission of the EC, 1978).

19 P. Schmitter, 'Reflections on the theory of Neo-Corporation', in G. Lehmbruch and P. Schmitter (eds.), *Patterns of corporatist policy-making* (London, Sage, 1982), pp. 4–5.

20 Bibes, *Europe elects its Parliament*, p. 69.

4
The development of the CAP

The policy-making process of the EC has created inertia in the development of agricultural policy both in its objectives and central provisions. But this should not divert attention from extensions to the CAP which have taken place from time to time. Before examining such policy measures, however, the argument for having any form of agricultural policy must be deployed.

The farm problem

Agriculture in the EC, as in other developed market economies, is faced by a number of difficulties which together are called 'the farm problem'. The persistence of relatively low farm incomes is widely acknowledged as the central issue, although the inefficiency of production on small farms and the long-run over-supply of markets are additional but inter-related features. We can view the farm problem in its many facets, therefore, as a gap between actuality and the values held by society for agriculture which governments try to meet by intervening in the economy.

Fundamental changes have taken place over the last few decades in the explanation offered by agricultural economists for the farm problem. Explanations have varied from an emphasis on the loss of foreign export markets to the inefficiency of marketing systems; from the technological revolution in farming to the structural rigidity of the agricultural industry. A consensus view on the farm problem has now emerged,[1] however, the central proposition of which is that agriculture is not master of the economic environment within which it operates. Summarising, aggregate demand for food in a developed economy increases only slowly. The income elasticity of demand for food declines with economic growth such that as real per capita incomes increase so the proportion of incomes spent on food declines. This is known to economists as Engel's Law. Income elasticities of demand approach or fall below zero, and the growth rate in the demand for food lies below that for the output of goods and services in the rest of the economy. Also, the population size of most developed countries has tended to grow only slowly, and recently has become static. Consequently, the relatively slow growth of demand for farm products requires agriculture to be a declining industry at an early point in economic development, at least as far as employment is concerned.

In addition, technological progress in agriculture tends to increase output at a rate greater than the increase in demand, and in a sense it is unfortunate that the introduction of the CAP coincided with a period of very rapid technological advance, particularly in the yield of crops per hectare. While individual farmers benefit by adopting new technology, in aggregate the industry tends to produce goods surplus to market requirements. Surpluses then depress agricultural prices with consequences for low farm incomes. The only long-term solution to the farm problem appears to be the transfer of resources, particularly labour, out of the industry and into more productive sectors of the economy. Those people remaining in agriculture can increase the size of both their farm businesses and their share of the aggregate agricultural income. This is termed the 'resource-adjustment' solution to the farm problem and is a long-term alternative to the short-term expedients of manipulating agricultural prices and restraining surplus production. The alternatives are often summarised as 'structural' and 'price policies' respectively.

For a variety of reasons, many related to the difficulties of obtaining alternative employment, farmers and farm workers have not left agriculture in sufficient numbers, nor at a sufficiently rapid rate, to provide a resource-adjustment solution to the farm problem.[2] Labour immobility is caused by features such as the elderly population age-structure of farming communities, the costs of geographical as well as occupational mobility, agricultural fundamentalist philosophies, and deficiencies in the rural educational[3] and employment infra-structure. Economic arguments concerned with asset fixity and the low 'salvage value' of human and physical resources employed in agricultural production have also been advanced.[4]

The importance of the contemporary explanation offered for the farm problem lies in the view of agriculture as a victim of general economic processes over which the industry has little control. In addition, the emphasis placed on the resource-adjustment solution has impressed upon society the realisation that if carried out in an uncontrolled or rapid fashion, the effect on the stability of rural areas, and whole regional economies, would be too costly certainly in human terms and probably in economic terms as well. Thus the CAP has been established both to manage the decline of the agricultural industry from its position as a large employer of labour, and to protect farm incomes while that transition is accomplished. For example, the number of people in the Nine who obtained their living from agriculture fell from nearly nineteen million in 1958 to under eight million in 1981. These figures are indicative of the social revolution experienced by rural areas in recent decades. In managing this revolution the EC has acted no differently from governments throughout the world, although arguably with rather less success.

A second reason for government intervention lies in the tendency for agricultural incomes to be more unstable from year to year than comparable non-farm incomes. The variability of weather conditions from year to year, or season to season, is a major contributary factor, and one over which agriculture again has little control. Prices can vary considerably in the face of weather-induced fluctuations in supply since the price elasticities of demand for agricultural products tend to be low in the short term. In general, relatively fixed quantities of foods are required by consumers irrespective of price. For farmers, price instability is a risk, the avoidance of which leads to the sub-optimal allocation of resources by inhibiting specialisation and the development of economies of scale in production. Also there are undesirable social consequences arising from price, and hence income variability, and they have been used to justify a wide range of policy measures designed to provide a more stable net farm income from year to year.

Phases of policy evolution

In retrospect the CAP can be seen as evolving in a cyclical fashion, a process that continues to operate at present. The pressure for change to policy measures builds up over a number of years, culminating in the introduction of new or revised legislation which often reflects the contemporary interpretation of the 'farm problem'. There follows a period of quiescence while the new measures are introduced and their effects monitored. But then follows mounting pressure for further change and the cycle is repeated. Six such phases can be recognised in the evolution of the CAP to date. Initially the emphasis was laid on farm prices and then successively on farm structure, farm income and regional disparities.

1957–62. Establishing the principles of the CAP
The complexity of the CAP can be attributed in part to the diversity of agriculture within the EC, and in part to the variety of national agricultural

policies previously in existence. The prior experience of the founding members of the Community with agricultural policy cannot be summarised in a few words, and interested readers should consult one of the extensive treatments given to the theme.[5] One feature, however, was a certain degree of commonality amongst the countries in that each had developed, by piecemeal action, a set of protective policy measures which was at once distinctive yet drawn from the same pool of available measures. The specific combination selected by each country can be attributed to the combined influence of six factors.[6] They are the stage of economic development, farm-size structure, resource endowment, population density, degree of food self-sufficiency, and the political influence of agriculture. West Germany, for example, had arguably the most developed economy and Italy the least; the farm-size structures of Belgium and West Germany were inferior to those of Denmark and France; France has possibly the best resource endowment of any country for agricultural production; population densities were significantly higher in Belgium and the Netherlands than elsewhere; the degree of self-sufficiency in major agricultural products was highest in France and lowest in Belgium and Luxembourg; the political significance of agriculture was highest in France. Not surprisingly the combination finally selected by the founder members of the EC comprised measures which were already in force in one or more countries, their kinship with minimum price systems and tariff barriers being particularly evident.[7] From the outset prior experience of protection from outside suppliers and high self-sufficiency tended to mould the main principles upon which the CAP was to be founded, rather than any rational conception of the welfare needs of the EC as a whole.

The first formal steps in the search for policy methods appropriate to the CAP were taken at the Stresa Conference of 1958. But not until 1960 did the Commission come forward with firm proposals, and it was August 1962 before the first regulations came into force.[8] The difficulty encountered in agreeing on even the general terms for a CAP is illustrated by the facts that it took forty-five meetings lasting 137 hours and 582,000 pages of documents to achieve.[9] Indeed, the negotiations leading to the Brussels Agreements provided, so far, the longest debate on agriculture in the EC, lasting from mid-December 1961 to 14th January 1962. For the first time the device of the 'stopped clock' was used enabling decisions taken after a specified date to be deemed to have been taken at the proper time. Finally, however, six basic principles for the CAP were agreed:
1 A single market area.
2 Free internal movement of agricultural products.

3 A uniform external tariff.

4 Common prices within the market for the main products.

5 Community preference in agricultural trade.

6 Sharing the financial burden of the CAP.

Often these principles are summarised under three headings: market unity, Community preference and financial solidarity.

The advantage of providing a common degree of protection to farmers was to be that it removed distortions in competition and would lead to an optimal location of production within the Community. Fair competition between members of the Community was adopted as the driving force of production rather than planning; but it was assumed, falsely as it transpired, that Member countries would follow broadly similar rates of economic growth and full monetary union would be achieved.

The measures agreed upon in the early 1960s remain fundmentally unaltered today. Generalising, the Community's internal agricultural markets are protected from outside competition through a system of import taxes (variable levies), while internal prices are maintained at annually adjusted minimum levels by unlimited 'intervention' in the market. EC agencies buy-up produce surplus to market requirements and either pay for their storage while alternative markets are found, or arrange export subsidies for their disposal through international trade. The intervention system is designed primarily for conditions where supply and demand are in broad equilibrium, and because of its 'open-ended' nature is unable to cope adequately with the long-term production of agricultural surpluses. Within the limits set by these arrangements, prices are freely derived by supply–demand conditions.

The agreements reached in 1962 also included the method of financing the CAP. Member States were to contribute to a European Agricultural Guidance and Guarantee Fund (FEOGA). The Guidance Section was established to provide finance for structural improvements in production and marketing, but it operated from the outset with a limited budget. The improvements financed by the Guidance Section have to be given additional support from national budgets while individual recipients of aid are also involved in expenditure. The Guarantee Section, by comparison, provided market support for a range of agricultural products. Expenditure was allowable on official intervention to support the market, and on restitutions or subsidies paid on exports of Community produce to enable them to be sold at lower world market prices. In the 1970s, Monetary Compensatory Amounts (MCAs)[10] were also paid from this Section of the Agriculture Fund.

The method of financing the Agriculture Fund has altered through time, but at present subventions are made to it from the general Community Budget.

1963–7. Implementing the principles of the CAP

A twelve year 'transition period' to a CAP had been agreed in the initial negotiations leading to the Rome Treaty. In May 1960 the completion date was advanced to 1966, although subsequently it fell back to 1968. By that time national agricultural policy measures were to be aligned to the CAP so as to eliminate any distortions to fair competition between producers in the Community. The first variable levies, however, were applied on July 30th 1962 covering imports of grain, poultry, eggs, and pigmeat, while certain quality standards were applied to fruit, vegetables and wine. In the following year, agreement in principle was reached on marketing arrangments for dairy products, beef, veal, rice and fats, and in 1966 for fruit, vegetables, sugar, and oils. By 1968 a single market had been achieved for most of the main agricultural products; tobacco and wine were added in 1970, hops in 1971, and sheepmeat in 1980 (Table 3). Some products, however, such as potatoes, are still without Community marketing regulations.

Common markets were established first and then common prices. Following agreement in 1964, the harmonisation of national product prices had its first major success in July 1967 with the application of common cereal prices, together with arrangements for eggs, top fruit, and pigmeat. Cereal production was identified as the key sector in early developments owing to its role both as a staple food and as a raw material for livestock production. Common prices for dairy products, beef, veal, rice, sugar, oilseeds, and some fruit and vegetables followed in 1968.

From the outset, there was a conflict of interest between France and West Germany in setting price levels since domestic prices had been higher in the latter country. French negotiators wished to avoid the inflationary effects of higher prices, while those from West Germany tried to minimise the adverse effects of lower prices on their agricultural industry. The compromise was political in nature. In some cases – cereals, sugar beet, and grain-fed livestock – the common price was approximately an average of the price levels existing in Member countries. For other products, especially dairy products, the compromise price lay at the upper end of the range of national prices. The consequences for any country varied from product to product. France, for example, benefited from higher prices for cereals and dairy products, but was adversely treated by the price levels set for fruit, vegetables, and eggs. Subsequently, farmers in Italy, Luxembourg and West Germany were compensated from the Agriculture Fund for the fall in their prices. These compensatory payments ceased in 1970 by which time nearly 414 m. Units of Account (UA)[11] had been expended, approximately 68 per cent accruing to West Germany.

In general, the compromise price levels were set substantially above those proposed by the Commission, especially in the case of dairy products, while intervention prices were provided for an unlimited volume of production. The supply-control measures previously practised in exporting countries, for example, were abandoned. Not only was little objective analysis given to setting the initial price differentials between products, but no mechanism for relating overall supply to demand was created. Nevertheless the CAP was endowed with a variety of methods for intervention in the markets of different products: support prices, for example, were established for 72 per cent of agricultural production covering items such as wheat, pigmeat, poultrymeat and eggs; deficiency payment systems of support were created for nearly 3 per cent of production, mainly involving durum wheat and olive oil; and aid at fixed rates was established for a few specialist crops such as cotton-seed, flax and hops.

1968–71. Negotiating a realignment of the CAP
Even as national marketing policies were being harmonised and common prices instituted, it became apparent that a price or commodity-based policy did not conform sufficiently to the realities of agricultural production in the EC. There was, in general, a growing recognition that price support is a very inadequate mechanism upon which to build an agricultural policy, since price controls distort the market mechanism for allocating resources. In addition, the large producers tend to benefit more than the small, farm rents and land values become inflated, and high prices stimulate possibly unwanted production. Surpluses in the EC were stimulated particularly in milk, sugar, and wheat. In 1969, for example, one sixth of the EC wheat crop had to be denatured in an attempt to balance supply and demand by diverting wheat from human to live-stock consumption. The pricing system made no concessions to the variety of agricultural systems practised on farms of different sizes and in different regions, even though Article 39 (2) specifies that regional disparities in structure and natural features should be taken account of in working out the CAP. Any policy designed for self-sufficiency, however, will tend to produce surpluses in the long run, and the method of funding the Community Budget itself puts pressure on all Member States to be as self-sufficient as possible so as to limit payments of import levies.

Table 3. A summary of price and marketing regulations under the CAP.

Product	Regulations	Product	Regulations
Common wheat	E, I, T, Th, V	Milk products:	
Durum wheat	E, I, P[a], T, Th	butter	E, I, Th, V
Barley	E, I, T, Th, V	SMP[c]	E, I, Th, V
Rye	E, I, T, Th, V	cheese[d]	E, I, Th, V
Maize	E, I, T, Th, V	Beef: live	G
Rice	E, I, T, Th, V	meat	C, E, I, V
Sugar:		Sheepmeat	B, E, I[e], R, VP[e]
white	E, I, T, Th, V	Pigmeat	B, E, I, S, Su
beet	M	Eggs	E, S, Su
Oilseeds:		Poultrymeat	E, S, Su
colza/rape/		Fresh fruit &	
sunflower	D, I, T	vegetables	B, C, E, R, W
soya/linseed	D, G	Live plants	C
castor	D, G, M	Olive oil	D[f], E, I, T, Th, P, V
cotton	P	Wine	C, E, G, I[g], R
Peas & field beans	D, M, T[b]	Hops	C, P
Dried fodder	D, P	Seeds for sowing	P, R[h]
Fibre flax and hemp	P	Tobacco	C, D, E, I, N

B: basic price; C: customs duty; D: deficiency payment; E: export refund; G: guide price; I: intervention price; M: minimum price; N: norm price; P: production aid; R: reference price; S: sluice-gate price; SU: supplementary levy; T: target price; TH: threshold price; V: variable levy; VP: variable premium; W: withdrawal price
a: certain regions only; b: activating price; c: skim milk powder; d: Italy only; e: alternative methods; f: consumer subsidy; g: storage contracts & distillation; h: hybrid maize only
Source: R. Fennell, *The Common Agricultural Policy of the European Community* (London, Granada, 1979), p. 106

Consequently the cost of supporting the production of agricultural surpluses through intervention buying became a major impetus to the realignment of the CAP.

Further, price policy was attempting to meet a second objective of the CAP in providing a fair standard of living for those employed in agriculture. It was found, in common with other developed countries, that price controls could not at the same time meet farm income and production objectives. Prices that were set so as to maintain a reasonable return to producers on small farms were so favourable as to encourage the production of surpluses from the industry in aggregate.

Four constraints operated to prevent the required internal price changes necessary to bring demand and supply for agricultural products into equilibrium, constraints which operate to the present. The first concerns existing suppliers of farm products. Any alteration in price relativities designed to suppress the output of one product, but with a compensating increase in the production of another, immediately conflicts with the interests of existing suppliers. Concern with the interests of developing countries, for example, in part explains the continuing importation of vegetable oils which Member countries could themselves produce with a shift in crop production. Secondly, quite large price changes are required to effect a substantial alteration in the balance of production, and such large changes are generally precluded by the farm-income role of price policy. Thirdly, there is a marked regional association of products within the EC and there are limits to the substitutability of certain crops. In Italy, for example, there would be a problem of replacing the regionally localised production of rice and tobacco. The fourth constraint on realigning price levels has been the political influence wielded by the agricultural industry. This influence has been exercised to prevent the disruptive effect on farm incomes and production patterns that a realignment of price levels would bring, and has included animal-feed firms, fertiliser manufacturers, farm machinery producers, seed merchants, transport contractors, slaughterhouse operators, dairy managers and meat processors.[12] The interests ranged against any fundamental changes in the CAP have been, and remain, formidable.

The response of policy makers to the farm income/surplus production conflict within the CAP was to recognise a duality in the agricultural industry. A commercial or production sector, composed of larger holdings, was distinguished from a non-commercial or small-farm sector. Price policies were to orientate production in the former sector, while 'structural' policy measures were to deal with the problem posed by the numerically more significant farms in the latter. This approach gave further support to the argument that an effective agricultural policy cannot treat all farms in the same way.

The policy measures took five years to negotiate, thus emphasising again the lengthy decision-making process of the EC when matters of a surpa-national character are at issue. The process was initiated in December 1968 by a Commission memorandum on the reform of agriculture in the EC.[13] This became known as the 'Mansholt Plan' after Sicco Mansholt, then the Commissioner for Agriculture. On prices, the Plan proposed a lowering of price levels over the medium or short term so as to reflect the supply–demand situation for each commodity. Prices were to reflect self-sufficiency not income levels, and production was to be limited by withdrawing five million hectares of land from agriculture and by a slaughter programme, with compensation, for dairy cows. On structure, the Plan proposed a more rapid reduction in the number of farms and farmers under a long-term programme of farm consolidation and amalgamation. The children of farm families were to be diverted into non-farm work, elderly farmers were to be compensated for retiring early from farming, younger farmers were to be retrained for non-farm employment, and finance was to be provided to modernise those farms remaining in production. The Plan also proposed that more encouragement be given to co-operation amongst smaller farms. The Vedel Report[14] on French agriculture, in August 1969, came to broadly the same conclusions, but proposed even greater reductions in the agricultural labour force and productive agricultural land.

Reaction from the agricultural industry to these radical proposals was understandably hostile, and fifteen months passed before a much diluted and revised memorandum was produced for Council debate. Final agreement on the policy proposals was not reached until March 1971, and then only after a marathon session of the Agriculture Council. The issuing of Directives was further delayed until April 1972. The new Directives harmonised existing national policy measures on farm structure, but each country retained the responsibility for developing its own range of programmes. Prior to the Directives the collective expenditure of Member States on social and structural measures had often exceeded that for market support, although national measures varied greatly. France and the Netherlands, for example, had established agencies to buy and sell land for farm consolidation and amalgamation; France, the Netherlands, West Germany, and Belgium had retirement pensions for farmers; and all countries except Belgium and the Netherlands had low-interest, capital loan schemes for farmers wishing to purchase land for amalgamation. The EC 'structural' measures were voluntary for the farming community and enabling for the Member countries. Directive 72/159/EEC assists farm modernisation; Directive 72/160/EEC encourages the retirement of older farmers from agriculture and the reallocation of land to the other farmers; Directive 72/161/EEC promotes the training of farmers and farm workers for employment within agriculture. National governments are empowered to vary the application of the Directives within national boundaries so as to take account of regional variations in the need for 'structural' policies, while the Agriculture Fund pays one quarter of the cost of national programmes.

1972–6. Implementing the extended provisions of the CAP
The implementation of the revised provisions of the CAP coincided with the enlargement of the Community in January 1973 from six to nine countries. The new members – Denmark, Ireland, and the United Kingdom – were

phased into the CAP over a five year transition period which ended in December 1977. Initially, their farm price levels were below those operating in the rest of the EC. 'Accession Compensatory Amounts' (ACAs), therefore, were necessary to act as an export subsidy for the Six but were counterbalanced by a levy on imports from the new Member countries. Year by year, price differences were reduced for most products, and for a time CAP and national measures ran side by side. Indeed, for potatoes, sheepmeat and wool, national measures, such as the deficiency payment system in the United Kingdom, were necessary in the absence of price support mechanism in the Community. The United Kingdom also obtained agreement to operate a special system of variable slaughter premiums to support the national beef sector.

Bearing in mind the difficulty with which the policy measures of the CAP had been agreed, it is hardly surprising that the new members were unable to effect any fundmental changes in agricultural policy in their negotiations for membership. Even the United Kingdom in 'renegotiating' her terms of entry in 1974 was unable to extract more than the most minor of concessions on her contributions to the Budget of the Community – the so-called 'financial mechanism'. Indeed, it was widely considered that the inclusion of the United Kingdom in the EC, with her relatively large requirement for imported food, would help resolve the dual problems of the CAP, namely endemic surplus production of certain commodities and a sharply rising budgetary cost.

Nevertheless, the new members ensured that the reappraisal of the CAP was continued. Thus a Memorandum on the 'Improvement of the CAP' was sent by the Commission to the Council in October 1973,[15] while work on a 'Stocktaking of the CAP' resulted in a Commission report early in 1975.[16] Taken together, the 1973 and 1975 reappraisals of the CAP concluded that there was a balance of benefits and shortcomings arising from the CAP. On the credit side the reports noted security of supply, the guarantee of reasonable prices to consumers, the stabilisation of markets and the increased productivity in agriculture. On the debit side were placed the short-term imbalance on certain markets such as beef and veal, the long term structural surplus of the milk sector, and the remaining lag of agricultural as compared with non-agricultural incomes. A number of recommendations were contained in the reports, although the system of price guarantees was to remain the central provision of the CAP. These recommendations included elimination of long-term differences in MCAs, the implementation of Directives on 'Less Favoured Areas', temporary direct aid measures, consumer subsidies, and international trade agreements.

At the same time as these reports, there emerged a growing interest in, and awareness of, the regional dimension of the 'farm problem', reflected, for example, in a 1973 analysis of agricultural incomes in the enlarged Community.[17] The increasing variability of farm incomes between regions gave impetus to a Directive on 'Mountain and hill farming in certain less-favoured areas'. The Directive was adopted by Council in January 1974, and was subsequently issued as Directive 75/268/EEC in February 1975. The Directive enables Member countries to define problem agricultural areas within their boundaries (Directives 75/269/EEC – 75/276/EEC). Three major types have emerged: mountain and hill areas with permanent handicaps to production caused by steepness of slope, altitude or soil type; areas with low population densities and severe depopulation; and areas with special handicaps or with poor infrastructure (Figure 4). Discriminatory financial assistance can be given to agriculture on the basis of location since it has an essentially social rather than economic objective. Assistance takes the form of an annual 'compensatory allowance' which is related to the land occupied and in part covers increased costs of production. In the United Kingdom, for example, it is paid as a grant per head of meat livestock (breeding ewes and beef cows) kept on each farm. In addition assistance can be provided to groups of farmers joining together for fodder production, while more favourable rates of aid are permitted for farm modernisation and investment in non-agricultural enterprises such as tourist facilities and rural crafts. These last measures give specific recognition to the role of non-farm activities in supporting the incomes of farm-families for the first time.

The measures contained in Directive 75/268 are of considerable significance: they permit price policy to be abandoned as the instrument of income support; the direct payments are selective to a defined stratum of farmers; they can be divorced from production. Only a quarter of the expenditure under the Directive is funded by the Guidance Section of FEOGA, and not all Member States have implemented every aspect of the Directive. For example, the United Kingdom has not implemented fully the non-agricultural aspects of the Directive dealing with rural crafts and tourist facilities.

The mid-1970s also saw a renewed if limited effort by the Commission to create a more rational set of price levels ('prudent' pricing – 1977) and price relativities for farm products within the Community. It reflected an emerging shift of emphasis for agricultural policy in the Community: the consumer, the price of food and the security of food supplies all gained in

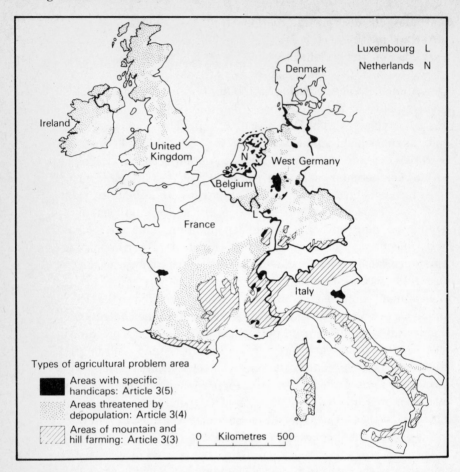

Figure 4. Areas eligible for assistance under the 'Less Favoured Areas' Directive

influence. With world-market food prices rising nearer to levels within the EC, there was for a time in the mid-1970s a possibility of realigning price levels. However, the inflationary pressures of the 1970s, fuelled by rising energy costs, focussed attention once more on the budgetary costs of the CAP, and on the demand by farmers that product prices keep pace with the cost of production. Once again, price agreements came to reflect the trade-offs of national interests rather than any 'Community interest'.

1977–80. Growing criticism of the fundamental provisions of the CAP
Criticism of the fundamental provisions of the CAP gathered momentum in the late 1970s. The central issue to emerge was the mounting financial

burden of existing policy measures which threatened to exhaust the established system of funding the Community Budget. Two particular weaknesses of the CAP had become exposed.

First, the open-ended character of the price support system through intervention buying proved costly when faced by the sustained production of agricultural surpluses. Corrective reductions in intervention prices had not been possible for a mixture of social and political reasons, and at best price levels were being set so as to allow inflationary forces to erode their real value.[18] The financial problems of the intervention system at times became so severe as to create a division between the Commission and the Council of Ministers. In 1979, for example, the Commission was obliged to disassociate itself from the Agriculture Ministers' refusal to control agricultural expenditure in the proposed Budget for 1980.

Secondly, increasingly complex arrangements had become necessary to maintain common or uniform institutional (CAP) prices within the EC as far as intra-Community trade was concerned. Throughout the 1960s the economic performance of Member States increasingly diverged culminating in the devaluation of the French franc and the revaluation of the West German mark in 1969. The system of MCAs, described more fully in the following chapter, was established to remove any advantages or disadvantages in trade accruing from alterations in monetary exchange rates. One important outcome was the failure to maintain common prices for producers within the EC while another was the creation of an at times substantial drain on the Community Budget. The formation of a new European Monetary System (EMS) in 1979, however, offered a fresh but uncertain prospect of a return to monetary stability within the EC.

Thus the existing provisions of the CAP as regards intervention buying and the maintenance of common prices were increasingly questioned, and a drawn-out debate over the nature of a reformed CAP ensued. While that continued, adjustments were made to existing policy measures in an attempt to meet some of the short-term financial pressures. The dairy sector, for example, was subjected to a variety of measures either limiting production of milk or increasing consumption by subsidising the price of dairy products to specified groups in society.[19] Not all Member States have operated every one of the measures nor have the measures been operative in a country the whole time. Similar schemes to limit the production of wine were introduced in 1976: Regulation 1162/76 prohibited new plantings of wine grape varieties for a number of years; Regulation 1163/76 allocated funds for producers who grubbed up unsuitable varieties.

Another area of criticism of the CAP was its failure to reduce regional variations in farm income levels within the EC. Following the precedent set by the LFA Directive of 1975, a number of regionally specific grant-aid schemes were established in the late 1970s. In 1978 and 1979, for example, programmes of drainage operations were approved for the west of Ireland (Directive 78/628/EEC) and the border between Ireland and Northern Ireland (Directive 79/197/EEC) respectively. Also in May 1978 a substantial allocation of funds was approved for producers in the Mediterranean regions of the EC (the Mediterranean package). The measures included investment aids for irrigation schemes in southern Italy (Regulation 1362/78) and Corsica (Directive 79/131/EEC), restructuring of vineyards in certain Mediterranean regions of France (Directive 78/627/EEC), and the improvement of social infrastructure in rural parts of Italy and southern France (roads, electricity, drinking water – Regulation 1760/78). Financial assistance was also approved for the marketing of olive oil, fruit and vegetables. In retrospect, these measures marked the first moves in a realignment of the CAP so as to improve the farm incomes of producers in southern parts of the Community. Such a realignment must also be seen in the perspective of Greece's membership of the Community from January 1981, and the prospective membership of Spain and Portugal in the mid-1980s. In the long-term, the gain of the Mediterranean regions will be at the expense of producers in the northern regions of the EC who historically have dominated agricultural policy making.

1981 – present. Gradual 'reform' of the CAP

'Reform' is probably too strong a term to describe the amendments being made to the CAP in the 1980s, although cumulatively such changes could have a reforming effect. A pivotal role has been played by a 1981 report from the Commission to the Council of Ministers on the restructuring of the Budget.[20] The report followed problems in negotiating a solution to the imbalance of the United Kingdom's net contribution to the Community's Budget in May 1980, and reflected a growing concern that the 'own resources' method of funding Community expenditure would soon be exhausted. The Commission responded to its 'mandate' by stressing the need to curb expenditure under the CAP.[21]

Two main thrusts of policy are being attempted. First, renewed attempts are being made to contain the price increases negotiated annually in the Council of Ministers, particularly cereal prices. Secondly, the mechanism of 'production' or 'guarantee' thresholds has been introduced, first for sugar and from 1982–3 for cereals (except drum wheat), milk, rapeseed and processed tomatoes. Prices will be adjusted downwards, or producers will be expected to contribute towards the cost of disposal of surpluses, if quantitative production targets are exceeded. On rapeseed, for example, the Council fixed a guarantee threshold of 2.15 million tonnes for 1982 and laid down that the intervention price for 1983 would be reduced by one per cent for every 50,000 tonnes by which the average production for year 1980–82 exceeded this threshold. Production in 1982 came to 2.66 million tonnes, giving a three year average of 2.2 million tonnes for the EC. Accordingly the 1983–4 price proposals included a downward adjustment of one per cent for rapeseed and a new guarantee threshold of 2.29 million tonnes. An even more stringent system of delivery quotas was introduced for milk in 1984. For individual farms or dairies a super-levy (75 per cent of the milk target price) will be attached to milk deliveries above those achieved in 1981 (plus 1 per cent).

It seems doubtful if such 'reforms' on their own will be sufficient in the long-term to preserve the present structure of the CAP. Much will depend on the relative levels of world and CAP prices for agricultural products and on the resulting financial burden for the Agriculture Fund. Equally important will be the degree of budgetary restructuring that proves possible, especially as regards the present limits placed on funds such as the one per cent VAT contribution from Member States. If the ceiling on the 'own resources' available to the Budget were to be lifted, the main pressure for continued 'reform' of the CAP would in large part be removed. The cost of the CAP, the structure of funding for the Budget and reform of the CAP are all inextricably interlinked.

Conclusion

A viable policy measure must have a purpose that moves with the grain of history, an operation that is manageable to those who administer it, acceptable to those who finance it, and tolerable to those who are affected by it, and proper timing in its implementation.[22] The CAP reflects several of these dictums. For example, when the policy has proved administratively difficult to operate, or unduly costly, so moves have been made to alter the CAP and make it more acceptable to farmers, consumers and administrators. The process of change, however, is slow and the present degree of criticism of the CAP suggests a failure in the 'proper timing' of revised policy measures.

The existing structure of the CAP also no longer 'moves with the grain of history'. Most agricultural policies, and EC agricultural policy has been no exception, have been broadly conceived as providing a framework for the solution of the 'farm problem': farm incomes have been supported while the number of people employed in agriculture has been reduced. Traditionally, the social evolution of rural areas has been at issue through the process of rural–urban migration. But with the exception of a few regions in the EC, the main phase in the transition of agriculture from a high to a low employer of labour is almost complete. The conventional age-retirement of farmers without replacement seems sufficient to continue the process of transition in most parts of the EC, but at a much reduced level. For those regions still with a sizeable farm population, however, rural–urban migration is no longer an appropriate solution to the 'farm problem'. On the one hand, urban-industrial centres now have their own long-term problems of unemployment and low incomes; on the other, the desirability of further concentrating the population into industrial complexes is increasingly questioned. Thus the provision of non-farm employment opportunities within regions having relatively large farm populations appears to be necessary for the 1980s, including the fostering of part-time farming. Such a view places agricultural policy firmly in the wider field of social and regional policy;[23] a 'regional rural development' dimension seems to be the next required phase in the evolution of the CAP.[24]

Moreover, reducing the size of the farm population has not resolved the problem of surplus agricultural production. Not only is the problem spreading to more farm products as levels of self-sufficiency rise, but available technology presages no let up in the rising productivity of the agricultural sector. A broadly applied system for containing over-production is required so as to render the CAP 'manageable'. Such a system will need to involve production targets and substantially reduced levels of price support. In introducing guarantee thresholds in 1981 the Commission has begun the necessary reorientation of the CAP. Nevertheless, such developments will need to be counterbalanced by a greater reliance on direct income supplements divorced from production for the small and medium sized farms in the EC. This aspect of policy making has at last been accepted by the Commission. However, since the need for direct income support, together with responsibility for the production of surpluses varies within the Community, a stronger regional dimension to the CAP can be justified.

Finally, emphasis has been laid on the widening scope of the CAP, yet the considerable influence that can still be exerted by a national government over its domestic agricultural industry should not be overlooked.[25] Some products are still not covered by Community marketing regulations; each country may determine the way in which a Directive is interpreted;

standards in administering the CAP may be varied – for example the quality of products taken into intervention; taxation, interest rates and national planning laws for agriculture may be altered; monetary compensatory amounts are influenced by government decisions on the exchange rate for the national currency. Consequently, nearly two-thirds of all expenditure on agriculture is still borne directly by national budgets rather than the Agriculture Fund, and this perspective on the CAP cannot be ignored in analysing the geography of agriculture under the CAP.

Notes

1 T. E. Josling, 'Agricultural policies in developed countries: a review', *Journal of Agricultural Economics*, XXV (1974), pp. 229–64.

2 G. James, *Agricultural policy in wealthy countries* (Sydney, Angus and Robertson, 1971), pp. 332–3.

3 G. S. Tolley, 'Management entry into U.S. agriculture', *American Journal of Agricultural Economics* LII (1970), pp. 485–93.

4 James, *Agricultural policy in wealthy countries*, pp. 333–4.

5 For example: M. Tracy, *Agriculture in Western Europe: challenge and response 1880–1980* (2nd ed.) (London, Granada, 1982); and OECD, *Trends in agricultural policies since 1955* (Paris, OECD, 1961).

6 T. Heidhues, *Agricultural policy: the conflict between needs and reality* (Gottingen, University of Gottingen, 1970).

7 Tracy, *Agriculture in Western Europe*, p.273.

8 J. Bourrinet, *Le problème agricole dans l'integration européenne* (Montpellier, 1964).

9 P. G. Minnemann, 'Agriculture in France and the EC', in S. N. Fisher (ed.), *France and the European Community* (Columbus, Ohio State University Press, 1964), pp. 83–4.

10 MCAs were first introduced in 1969 to meet the effects of the changed valuations of the French and West German currencies: they are treated more fully in Chapter 5.

11 The unit of account (UA) was initially based on the unweighted average value of the national currencies and thus tended to reflect the influence of the strongest currencies in the EC. In 1960 1 UA = 0.88867088 grammes of fine gold = 1 $ US.

12 E. Deakins, *EEC problems for British agriculture*, Fabian tract 408 (London, Fabian Society, 1971).

13 Commission des Communautés Européennes, *Memorandum sur la réform de l'agriculture dans la Communauté Economique Européenne*, COM (68) 1000 (Bruxelles, La Commission, 1968).

14 Ministère de l'agriculture, 'Rapport général de la Commission sur l'avenir à long terme de l'agriculture française 1968–85' (président M. G. Vedel), in *Le Plan Mansholt – Le rapport Vedel* (Paris, SECLAF, 1969), pp. 521–68.

15 Commission of the European Communities, *Improvement of the Common Agricultural Policy,* COM (73) 1850 final (Brussels, The Commission, 1973).

16 Commission of the European Communities, *Stocktaking of the Common Agricultural Policy*, COM (75) 1100 (Brussels, The Commission, 1975).

17 Commission of the European Communities, *Agricultural incomes in the enlarged Community* (Brussels, The Commission, 1973).

18 R. G. Harper, 'Criticism of EC farm price policy gains ground in West Germany', *Foreign Agriculture* XI (1973), pp. 2–4.

19 R. Fennell, *The Common Agricultural Policy of the European Community* (London, Granada, 1979), pp. 191–2.

20 Commission of the European Communities, *Commission Report on the Mandate of 30 May 1980*, COM (81) 300 final (Brussels, The Commission, 1981).

21 Commission of the European Communities, *Guidelines for European agriculture*, COM (81) 608 final (Brussels, The Commission, 1981).

22 R. E. Neustadt, *Presidential power: the politics of leadership* (New York, Wiley, 1960), p. 184.

23 A. Cairncross *et al*, *Economic policy for the European Community: the way forward* (London, Macmillan, 1974).

24 The first formal move in this direction has been provided by Commission of the European Communities, *The Commission's proposals for the integrated Mediterranean programmes*, COM (83) 24 (Brussels, The Commission, 1983).

25 J. S. Marsh, *U.K. agricultural policy within the EC*, Centre for Agricultural Studies Paper 1 (University of Reading, 1977), pp. 51–2.

5
Technical aspects of the CAP

There can be no disguising the sheer volume and technical complexity of the measures that comprise the CAP. The policy measures or instruments must at the same time protect the agricultural sector against foreign competition, offer price stability and reasonable economic returns to domestic producers, yet contend with the objective of free competition within the Community in the production and trade of agricultural products. The resulting variety of policy measures has been described and analysed in many publications,[1] and here only a summary of their main features is provided. Attention is focussed on how the CAP is financed and on the price and market mechanisms for seven of the major crop and livestock products of the EC.

The Budget
The EC is financed through a common Budget and the procedure which establishes it in any year (Articles 202–5 of the Rome Treaty) is long and the outcome uncertain. The financial year of the Community runs from January to December, and each Spring estimates for the following year are prepared by each institution in the EC and consolidated by the Commission. Estimates include the administrative costs of the various Community institutions together with the costs of the Community policies. The preliminary draft Budget is passed to the Council by 1st September at the latest. The Council establishes the draft Budget and forwards it to Parliament not later than 5th October (Figure 2). Agricultural interests in the Parliament then exert their influence in the ways previously described. The Parliament may approve the draft Budget or propose amendments either by increasing certain heads of expenditure subject to a given ceiling or by reallocating expenditure within the Budget.

Since 1970 the Budget has been divided into obligatory or compulsory

Table 4. Scaled national contributions to the EC Budget (% total)

Country	1962/3 -1964/5[a]	1965/6 -1966/7[a] (average)	1967/8 -1970/1[b] (average)	1971/2 -1972/3[c]	1973/4 -1978/9[cd] (range)
West Germany	28.00	31.25	31.50	32.90	25.53–32.13
France	28.00	30.92	30.00	32.60	22.99–25.30
United Kingdom	–	–	–	–	16.90–20.04
Italy	28.00	20.00	20.90	20.20	13.20–15.60
Netherlands	7.90	9.66	9.30	7.30	5.26– 6.00
Belgium	7.90	7.87	8.20	6.80	4.11– 5.28
Denmark	–	–	–	–	2.40– 2.58
Ireland	–	–	–	–	0.60– 0.71
Luxembourg	0.2	0.22	0.2	0.2	0.15– 0.17

a: 80–90 per cent on a fixed scale, 20–10 per cent according to each country's share of EC net imports (excludes the European Social Fund); b: 90 per cent of receipts from agricultural levies and customs duties on agricultural produce, the balance according to the schedule of contributions; c: all the receipts from agricultural levies and all CET (the proportion rising from 50 per cent in 1971 to 100 per cent in 1975), the balance according to the scale of contributions; d: from 1979 the balance to be made up by VAT contributions
Source: M. Butterwick and E. N. Rolfe, *Food, farming and the Common Market*, (Oxford University Press, 1968); European Communities *The Budget and the British Connection* (Brussels, 1979)

(three-quarters of the total) and non-obligatory (non-compulsory) expenditure. On the details of the first – mostly agricultural – the Parliament has the right to propose amendments but the Council has the final say; on the second, Parliament has the last word. Obligatory items are those which necessarily arise from the Treaty of Rome, such as from the CAP, whereas items such as the Regional Fund (since 1978) fall into the second category of expenditure. There has been no clear-cut division of expenditure into the two categories, however, and the allocation of expenditure is a continuing source of disagreement and bargaining between the Council and the Parliament. If, after the first reading of the draft Budget, the Council wishes to depart from Parliament's modifications and amendments, a formal 'conciliation procedure' is implemented in an attempt to obtain agreement. This often becomes necessary since the Budget emerges as the by-product of individual decisions taken in the various Councils. The Agriculture Ministers dominate the final outcome because their expenditure tends to squeeze out other spending.[2] At the second reading of the Budget in Parliament, around December each year, MEPs have no power to amend any changes that have been made to their proposals in the field of obligatory expenditure. They can, however, re-establish their amendments to the non-obligatory element and adopt the Budget. Alternatively, by two-thirds of the votes and a majority of MEPs, the Parliament can reject the whole draft Budget and ask the Commission to prepare a new draft. The powers of the European Parliament were strengthened in 1976 by the formation of a 'Public Accounts Committee' within the Budget Committee. In addition, a Court of Auditors has, from 1977, ensured that money voted by Parliament and Council is actually spent on matters for which it was allocated.

Member States have always paid a varying share of the Budget. Prior to 1971 the cost was met by a combination of receipts from agricultural levies on certain imports, an internal levy on EC sugar production, and national contributions on a fixed scale approximately in proportion to the GNP of each country (Table 4). Gradually an 'own resources' system of funding the Budget was introduced, and since 1979 the EC has automatically received all

Table 5. EC Budget expenditure 1972–83 (% total appropriations for payments)

	1972	1975	1978	1981	1983
By Sector					
Agriculture[a]	67.67	72.87	73.87	68.87	65.68
Regional	–	–	4.25	9.18	9.50
Social	2.33	6.39	4.52	3.57	5.96
Staff	4.49	3.45	3.13	3.49	2.85
Remainder	25.51	17.29	14.23	14.89	16.01
By Institution					
Commission	99.00	98.26	98.20	98.17	98.53
European Parliament	0.35	0.71	0.80	1.03	
Council	0.55	0.86[b]	0.80	0.59	1.47
Court of Justice	0.10	0.16	0.10	0.13	
Court of Auditors	–	0.01	0.10	0.08	
Receipts by Heading					
VAT element	–	–	43.10	53.90	55.36
Customs duties	45.96	57.31	39.00	32.50	28.24
Levies	–	9.10	16.70	11.70	9.28
Other	54.04[c]	33.59[d]	1.10	0.90	7.12
Total	4177	5825	12362	19328	26533
	m.UA	m.UA	m.EUA	m.ECU	m.ECU

a: maximum 84.3% in 1973; b: including Economic and Social Committee; c: of which national contributions 53.2%; d: national contributions 32.6%
Source: European Documentation, *The European Community's Budget*, Office for Official Publications of the European Communities, Brussels, Periodical 1/79 (1979), pp. 20–2, and 5/81 (1981), p. 19, pp. 44–5; *Commission of the European Communities, Seventeenth General Report on the activities of the European Communities* 1983, Brussels (1984), pp. 48, 54

the levies from agricultural trade with non-member countries, the internal sugar levy, all common customs duties (Common External Tariff – CET) and, to balance the Budget, up to one per cent of the value added tax (VAT) revenue of Member States (Table 5).

Clearly this method of funding the Budget places the greatest demands on countries such as the United Kingdom and West Germany which import a high proportion of their food requirements and/or have a high VAT revenue. Expenditure from the Budget, however, is dominated by the Agriculture Fund. Until recently over seventy per cent of each year's Budget has been expended on the Agriculture sector (Table 5), and the major recipients from the Budget, therefore, tend to be those countries with the largest agricultural production such as France and the Netherlands. As a result of the net balance of Budget contributions and receipts, a transfer of resources occurs between Members of the Community.[3] Calculations of the size of the transfers have proved very contentious and complex, not least because of the need to take account of MCAs and the trade or balance of payments effect of importing food at prices above the world market level. Looking just at the Agriculture sector of the Budget, however, net FEOGA receipts and the 'food trade' effect appear to benefit the Netherlands, Denmark and France at the expense of Italy, the United Kingdom and West Germany. The absolute magnitude of the transfers varies through time, but broader calculations for the whole Budget have revealed a similar pattern of costs and benefits (Table 6).

There is now a widely acknowledged inequity in the direct net contributions made by individual countries to the Budget. In the late 1970s, for example, the United Kingdom became one of the largest net contributors despite having the third lowest GDP per capita in the Community. Until recently, however, the redistributive or net cash transfer character of the Budget was not explicitly or formally discussed within the EC: the Commission resisted the notion of a *juste retour* when developing new policies with budgetary consequences; countries enjoying a net gain from the budgetary arrangements were, and still are understandably reluctant to change the *status quo* – any move to reduce the net contribution of one country requires a counterbalancing net increase in contributions from other Members of the Community. Even so, pressure from the United Kingdom to reduce the size of the country's net contribution to the Budget became severe in 1980. A temporary repayment mechanism was negotiated which amounted to approximately £2000m. ultimately spread over three years and gave formal recognition to the problem of net resource transfers within the EC. At the time of writing, however, the United Kingdom is still dependent

Table 6. Net receipts from FEOGA and the food trade effect of the CAP (average of 1977 and 1978 in £m)

Country	FEOGA account (net)	Food trade balance[a]	Net total cost (−) or benefit (+)[b]
France	+519	+84	+603 (+734)
Netherlands	+575	+248	+823 (+631)
Denmark	+371	+242	+613 (+618)
Ireland	+176	+289	+465 (+475)
Belgium/Luxembourg	+10	−33	−23 (+156)
Italy	−209	−538	−747 (−646)
West Germany	−295	−395	−690 (−671)
United Kingdom	−572	−138	−710 (−1123)

a: these are estimates based on certain assumptions and give only an indication of the orders of magnitude; b: figures in brackets are estimates for the whole Budget made by W. Godley and reported in M. Whitby, *The net cost and benefit of EEC membership*, Seminar paper 7, (Centre for European Agricultural Studies, Wye College, University of London, 1979), p. 31
Source: C. W. Capstick, 'The Common Agricultural Policy and recent developments in EEC agriculture, *Journal of University upon Tyne Agricultural Society*, XXVII, (1980), pp. 38–50

on an annually negotiated repayment to reduce the size of the net contribution to the Budget.

'National shares' from the Budget have undoubtedly influenced national attitudes to many agricultural proposals in the past and impinge directly on the CAP at present. For example, West German opposition to the structural reform proposals of the Mansholt Plan was based in part on the high proportion of the budgetary cost that the country would have to bear. Similarly, the United Kingdom's refusal to devalue the 'Green Pound' in 1976 was aimed at 'clawing back' what was seen as an unfairly high contribution to the Budget. At present disagreement on 'national shares' overflows into debates on the level of funding for the Budget, how the CAP should be developed and the annual EC price negotiations. All of these issues have net resource transfer implications and tend to be linked together in any political negotiations.

Paradoxically, though, the Budget of the EC has not had the same broad political or economic significance as national proposals for raising revenue and spending money. It has had no general effect on the level of employment or income within the EC, for instance. In part this arises from constraints placed on the Budget – it has a fixed if variable revenue and therefore must be retained in balance. In part the explanation lies in the small size of the

Budget compared with the aggregate of national expenditures. The 1978 Budget, for example, came to over twelve million EUA,[4] equivalent to £8.25b, but represented only 2.6 per cent of the combined national budgets of the Nine, or 0.8 per cent of the Community's GDP. National expenditures on EC policies, however, are not included in these figures.

Expenditure on the CAP has to be kept within the limits of the resources available to the Budget and hence FEOGA. Historically, the Guarantee Section has absorbed a high and increasing proportion of the Agriculture Fund (Table 7). This reflects the preoccupation of policy makers with the pricing and marketing of agricultural products, the limits placed on funds available to the Guidance Section (285 m UA per year for the period 1967–73, 325 m UA per year for the period 1973–80, 720 m EUA per year from 1980), and the continuing reliance of the CAP on national programmes to finance structural adjustments in agriculture.

Price and marketing mechanisms
An 'objective method' has been adopted by the Commission for determining the annual CAP price increases which would be necessary to give economically viable farms (based on a sample of 'reference farms') an increase in net income comparable with non-farm incomes.[5] Year-to-year

Table 7. Expenditure from FEOGA, 1965–83 (m. UA)

Year	Guarantee Section	Guidance Section	Guarantee Expenditure as % of Total FEOGA Budget
1965	77	25.6	75.0
1967	403	134	75.0
1969	2058	496a	80.5
1971	2727	757	78.3
1973	3806	350	91.6
1975	3980	325	92.5
1977b	6830.4	296.7	95.8
1979b	10440.7	403.4	96.3
1981b	11570.5	730.6	94.1
1982b	12405.6	650.0	95.0
1983b	15848.1	620.8	96.2

a: including additional expenditure to reduce the effects of grain price alignments; b: m. ECU
Source: European Documentation, *The European Community's Budget*, Office for Official Publications of the European Communities, Brussels, Periodical 5/1981 (1981), pp. 31–2; Commission of the European Communities, *The Agricultural Situation in the Community*, 1983 Report, Brussels (1984), p. 154

changes in farm costs and the movement of non-farm earnings are combined to calculate the 'necessary' change in farm prices, assuming that the volume and pattern of output remains unchanged. Of course, a complication is introduced when prices are set to induce a desired reorientation of production within the Community. In addition, the price set for each product takes into account the market situation and also the inter-related nature of the production of certain commodities – for example beef and milk. Finally, a 1.5 per cent deduction is made to account for the increased productivity which is assumed to result each year from 'bio-technical' progress.

Initially, the Commission consults the Management and Advisory Committees and hears representations on the draft proposals from agricultural and other pressure groups. Often conflicting results from, and interpretations of the 'objective method' are advanced since movements in national currency exchange rates can be subject to varying interpretations. Draft Regulations are placed before the Council early in each calendar year, but negotiations by the Agriculture Ministers can last until April or May before agreement is reached on a composite package of price changes. In most years such bargaining leads to higher increases in farm prices than initially proposed by the Commission (Table 8). These price levels then incur high expenditures on market support from FEOGA. Between 1974 and 1979, for example, expenditure increased by an average of 22 per cent each year. Despite more moderate price increases in recent years, especially when inflation is accounted for (Table 9), the limits of the 'own resources' system of financing the Budget have now been reached. As previously discussed, revised budgetary arrangements are having to include mechanisms for containing the cost of the agricultural price support system within FEOGA.

The complex task of translating the political decisions into product prices and regulations is undertaken by the Commission but can take weeks or even months to finalise. For crop farmers in particular, production decisions in any year have to be taken before agreement has been reached on institutional prices for that year. Political uncertainty, therefore, has to be added to the list of other features, such as weather conditions and market instability, with which agricultural producers must contend under the CAP.

The price levels arranged through the Commission and the Council apply to and vary amongst individual agricultural products (Table 8). Each product has a distinctive price and marketing mechanism, and a potentially bewildering array of technical terms has been devised to describe the pricing systems (Table 3). Regulations for cereals were the first to be established (1967), however, and they set the pattern for other products.

Table 8A. Agricultural price changes for 1980/1 and 1981/2 (EUA %)

Selected products and type of CAP prices	Price changes for 1980/1		Price changes for 1981/2	
	proposed	final	proposed	final
Pigment: Basic	+3.0	+5.5	+9.0	+11.0
Wine: Guide	+3.0	+5.5	+10.0	+10.0
Olive oil: Intervention	+1.5	+4.0	+6.0	+9.0
Milk: Target	+1.5	+4.0	+6.0	+9.0
Butter: Intervention	0	+2.3	+6.0	+9.0
White sugar: Intervention	+2.8	+5.3	+7.5	+8.5
Beef: Intervention	+1.5	+4.0	+6.0	+7.5
Sheepmeat: Basic[a]	–	–	+6.0	+7.5
Common wheat: Intervention	+2.0	+4.5	+6.0	+6.0
Average of all prices	+2.4	+4.8	+7.8	+9.2

B. Average price changes for 1981/2 in national currencies[b] (current % inflation rate in brackets)

Italy	+16.2 (21.0)	Netherlands	+10.4 (7.1)
Ireland	+13.9 (18.2)	Belgium	+10.4 (7.0)
France	+12.4 (12.9)	United Kingdom	+ 9.6 (13.0)
Denmark	+12.4 (10.7)	West Germany	+ 4.8 (5.7)
Luxembourg	+10.4 (7.3)	Average EC	+11.3 (10.6)

a: sheepmeat regulations operated from October 1980; b: includes changes in 'green' rates

Source: *Agra Europe* (various years)

Table 9. Average annual CAP price changes 1974–83 (EUA %)

Year	Commission proposal	Council of Ministers final price	Annual rate of change of farmgate/ input price ratio
1974/5	+7.2	+9.0	−11.3
1975/6	+9.0	+9.6	+4.3
1976/7	+7.5	+7.7	+3.4
1977/8	+3.0	+3.9	−2.4
1978/9	+2.0	+2.1	+1.2
1979/0	0	+1.3	−2.3
1980/1	+2.4	+4.8	−5.9
1981/2	+7.8	+9.2	−2.5
1982/3	+9.0	+10.4	−0.1
1983/4	+5.5	+4.2	n.a.

Source: *Agra Europe* (various years)

Cereals: The basic principles of all the marketing schemes are to allow price formation to be determined by the free interplay of competition between producers within the Community, but to maintain a common, minimum level of prices, together with protection from non-EC producers, so as to yield reasonable and stable incomes to the farm sector. To these ends *target* prices are set for the internal market, in the case of cereals for durum wheat, common wheat, barley, rye, and maize. The target price is established for Duisburg in West Germany as the location in the EC with the largest grain deficit. The price represents the return it is hoped producers will achieve on the open market when their crops are delivered to a merchant; it is a wholesale not a farm-gate price.

A floor to the internal market is supplied by an *intervention* price (Figure 5). A single intervention price is fixed for common wheat, barley and maize, whereas rye and durum wheat have individual prices. Intervention prices are set for the Ormes Intervention Centre in France which is the location in the EC having the greatest surplus for all cereals at the wholesale level. Ormes, however, is but one in a network of intervention centres located throughout the EC. National intervention agencies operate through the centres to support and manage the market by purchasing, storing, and reselling products offered to them at the common intervention price. Usually, certain quality and quantity standards have to be met before surplus agricultural products can be taken into intervention and prices are often varied, monthly in the case of cereals, to allow for farm-storage costs during the year. This form of pricing encourages orderly marketing. In some countries existing bodies have been adapted to carry out the necessary administrative functions, in others new bodies have been created. A useful if broad distinction can be drawn between general purpose agencies and commodity-centred agencies.[6] Examples of the former can be found in West Germany where intervention for all commodities is carried out by the *Bundesanstalt-fürlandwirtschaftliche Marktordnung* in Frankfurt, while export refunds are handled by the *Oberfinanzdirektion-Hamburg*. In the UK there is only one administrative organisation: the newly created Intervention Board for Agricultural Produce (IBAP). Commodity handling functions are delegated to other pre-existing bodies acting on behalf of the IBAP: the Home Grown Cereals Authority (HGCA) deals with all aspects of cereals intervention; the Meat and Livestock Commission (MLC) handles support arrangements for sheep and beef; the Agriculture Departments of England and Wales, Scotland, and Northern Ireland oversee matters such as the volume and quality of commodities held in intervention stores.

A structure of commodity-centred agencies has been preferred in France and the Netherlands. In France, for example, the *Office National Interprofessionnel des Céréales* (ONIC) handles cereals and the *Société Interprofessionnelle des Oléagineux* (SIDO) oil seeds. The comparable agencies in the Netherlands are *Hoofdproduktschap voor Akkerbouprodukten* and *Produktschap voor Margarine, Vetten en Olien*. Nevertheless these last two agencies do not cover buying-in operations which are handled for all sectors by the *Voedselvoorzieningsin-en Verkoopbureau* (VIB).

For most products, EC price levels are above those at which agricultural commodities can be traded on the world market. As a means of protection against lower priced imports, a *threshold* price is set and, in the case of cereals, is so calculated that when imported grain is landed at the port of Rotterdam and transported to Duisburg it will sell at no less than the target price. The threshold price for each cereal is the same at all ports but is increased seasonally to reflect the trend of farm storage costs for the domestic crop over the year. Grain imported into the EC below the threshold price is subject to a *variable levy* (Figure 5), the import price and levy being calculated and reset each day by the Commission on the basis of the cheapest import price (c.i.f.) of grain at Rotterdam. Grain exports from the EC, by contrast, attract export refunds (restitutions) which vary with the transport costs to the country of destination and the difference between the EC and world market price. An export levy may be imposed when world prices are higher than Community prices, as in 1973–4 when a world shortage of cereals occurred. Lastly, producers of durum wheat (used for pasta) in specified regions of Italy, where low crop yields prevail, qualify for a special flat-rate payment per hectare harvested.

Milk and milk products

The dairy sector also enjoys price support through target, intervention and threshold prices. The target price relates to milk of 3.7 per cent fat content delivered to a dairy. Since less than twenty per cent of milk sales in the EC are for fluid consumption, however, the market is supported by individual intervention prices for milk products such as butter, skim milk powder and certain types of Italian cheese. Threshold prices for imports apply to a dozen 'pilot' dairy products such as whey powder and Gouda cheese.[7] Prices vary in proportion to the amount of milk that is needed to produce a given amount of each product. In theory, trade in liquid milk between Member States is possible; in practice the imposition of national health regulations has been used to protect the dairy sectors of individual Member States. The consumer

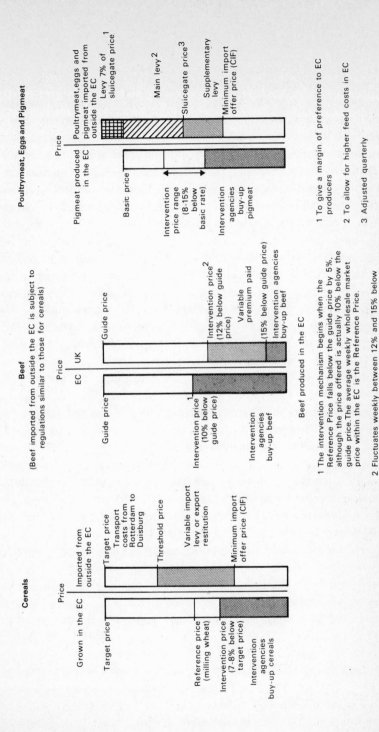

Cereals

Price

Grown in the EC

Target price

Reference price (milling wheat)

Intervention price (7-8% below target price)

Intervention agencies buy-up cereals

Price

Imported from outside the EC

Target price

Transport costs from Rotterdam to Duisburg

Threshold price

Variable import levy or export restitution

Minimum import offer price (CIF)

Beef
(Beef imported from outside the EC is subject to regulations similar to those for cereals)

Price

EC | UK

Guide price

Intervention price[2] (12% below guide price)

Variable premium paid

(15% below guide price) Intervention agencies buy-up beef

Price

Guide price

Intervention price[1] (10% below guide price)

Intervention agencies buy-up beef

Beef produced in the EC

1 The intervention mechanism begins when the Reference Price falls below the guide price by 5%, although the price offered is actually 10% below the guide price. The average weekly wholesale market price within the EC is the Reference Price.

2 Fluctuates weekly between 12% and 15% below guide price.

Poultrymeat, Eggs and Pigmeat

Price

Poultrymeat, eggs and pigmeat imported from outside the EC

Levy 7% of sluicegate price[1]

Main levy[2]

Sluicegate price[3]

Supplementary levy

Minimum import offer price (CIF)

Price

Pigmeat produced in the EC

Basic price

Intervention price range (8-15% below basic rate)

Intervention agencies buy-up pigmeat

1 To give a margin of preference to EC producers

2 To allow for higher feed costs in EC

3 Adjusted quarterly

Figure 5. Price support mechanisms under the CAP

price for fluid milk is determined nationally.

The dairy sector, because of its endemic surplus production, is also subject to a number of special measures which aim to either reduce supply or increase demand. Dating mainly from 1977, these have included a *co-responsibility levy* on milk production,[8] assistance to encourage small-scale producers to give up dairying,[9] premiums for the non-marketing of milk and milk products (ended in 1980), a consumer subsidy on butter, subsidies for the consumption of fluid milk in schools and butter by the armed forces, non-profit making organisations and people on social assistance, and disposal subsidies to users of skim milk or skim milk powder (SMP) as an animal feed.

Poultrymeat and eggs:

The market is supported by the regulation of imports, domestic producers being constrained only by strictly enforced quality standards. Every quarter the Commission fixes minimum import or *sluice-gate* prices. They roughly equal the world market value of the cereals required to produce one kilogram of poultrymeat or one kilogram of eggs in shell, together with the cost of other protein feed and overhead, production and marketing costs. Imports at below the sluice-gate price are subject to a *supplementary levy* (Figure 5). In addition a *main (basic) levy* is charged to offset the higher feed grain prices operating within the EC, official conversion rates for grain into eggs or poultrymeat being used. Lastly, a 7 per cent levy of the average of the sluice-gate prices of the previous year is imposed on imports to give EC producers a preference in the market.

Pigmeat:

The import control mechanism for pigmeat is similar to that for poultrymeat. In addition, the internal market price is supported by a *variable intervention* price. If the average market price (*reference* price) for pigmeat recorded in a selection of representative markets throughout the EC falls below 103 per cent of a *basic* price (Figure 5), intervention measures may be implemented at a price between 85 and 92 per cent of the basic price depending on market conditions. In addition, storage subsidies are commonly paid to processors and traders to hold meat off the market during serious gluts, while export restitutions have been widely used to disperse surpluses on the international market.

Beef and veal:

Each year the Council fixes a *guide* price for live mature cattle and calves which is the price that should be realised on the market prior to slaughter (Figure 5). The actual weekly market (*reference*) price is calculated as an average weighted price from representative wholesale markets throughout the Community.[10] If the reference price falls below 93 per cent of the guide price the intervention mechanism, mainly for animal carcasses, is activated throughout the EC, though purchases are made at prices based on 90 per cent of the guide price.[11] The beef market is protected by a customs duty and a variable levy. The proportion of the levy that is applied varies with different types of meat and the state of the market in the EC as indicated by the relationship between the reference and guide prices. For example, when the reference price is above 106 per cent of the guide price, no levy is charged.

A variant of the support system was introduced in 1974, largely at the instigation of the United Kingdom, and to date has been applied only in that country. Producers receive a *slaughter* or *variable premium* equal to the difference between the national reference price for mature cattle and an intervention price (called a target price in the U.K.) set weekly to encourage orderly marketing. Intervention buying only operates when the reference price falls below the guide price by between 12 and 15 per cent and is an amount larger than the allowable variable premium (Figure 5).[12]

Sugar:

Uniquely, the production of sugar from sugar beet is controlled in the EC by a system of production quotas, intervention buying and financial levies on sugar processors. A *basic* or *'A' quota* for white sugar is fixed annually for the whole Community and is supported by an intervention price. A smaller *'B' quota* is established at some proportion of the 'A' quota (23.5 per cent in 1981, for example) but is supported at 70 per cent of the full intervention price. Both quotas are divided amongst Member countries which in turn allocate quotas to individual sugar processors. Finally, quotas are distributed to the individual farmers who grow sugar beet under contract to each processing factory.

The intervention price for sugar is set at approximately 95 per cent of a target price for the main area of surplus production in northern France. Derived intervention prices are then fixed for other areas of production in the EC in relation to expected regional variations in supply and demand for sugar. Sugar processors are required to purchase sugar beet at or above a fixed *minimum price* for 'A' and 'B' quota production with allowance for variations in sugar yields and delivery charges to the factory. The processors, however, pay a levy on each tonne of sugar sold. The levy is repaid to those processors who store sugar during the year and which acts as an incentive for the orderly marketing of the commodity. In addition, a production levy is paid on 'B' quota sugar (extended to 'A' quota in 1981) at a rate not exceeding 30 per cent of the intervention price. This levy offsets the costs of export subsidies on sugar. Production outside the quotas ('*C*' *category* sugar) receives no official price support or export refund and must be sold outside the Community.

The internal market is protected by threshold prices and variable levies on white sugar, raw sugar, and molasses. The threshold price of white sugar, for example, is set at a level equivalent to the target price plus the storage levy and the transport cost between northern France (location of maximum surplus) and Palermo in Sicily (location of maximum deficit).[13] These arrangements are complicated by import agreements with certain developing countries whose economies depend largely on the production of cane sugar (Lomé Convention), and by market support under the CAP for isoglucose (maize sugar).

Common agricultural prices
Ideally, the EC acting as a common market should remove internal barriers to trade and so foster inter-regional competition in agricultural production. Relocation of production would then evolve in accordance with the principle of comparative economic advantage of the various regions. A prerequisite for such developments, however, is the achievement of common institutional price levels throughout the Community (target, intervention, threshold, guide, basic, sluice-gate prices). Common prices would give effect to the production-cost advantages of each region be they based on natural, structural or technological factors.[14] Common agricultural prices, however, have been in force in the EC only between 1967 and 1969. Prior to 1967 the common organisation of markets had not been achieved. After 1969 economic divergence between members of the EC severely disrupted the application of common institutional price levels.

In the 1960s each national currency in the EC had a fixed exchange rate against the US dollar. Such stable conditions permitted an Agricultural Unit of Account (AUA) to be established (1962) having the same value as the budgetary Unit of Account. Stable 'green' rates of exchange between the AUA and each national currency allowed common EC price levels, stated in AUA, to be converted into national currency equivalents. The original 'green' rates were the same as the market rates of exchange for each national currency. In 1969, however, France devalued the franc and West Germany revalued the mark. Both countries declined to alter their 'green' rates of exchange accordingly because of the prospective impact on national farm incomes and inflation rates. As a temporary measure to protect the EC intervention system and prevent trade distortions from speculation and profiteering, the system of MCAs was introduced.[15] MCAs acted as border levies and subsidies on trade in agricultural products equal to the percentage difference between the 'green' rate of exchange and the new market rate of exchange (monetary percentage) multiplied by the intervention price of the commodity being traded.[16] Thus, for France, MCAs were applied as a levy to raise the value of (cheaper) exports but were operated as a subsidy to decrease the value of (dearer) imported agricultural products. MCAs in West Germany, by contrast, reduced the price of (dearer) exports while raising the price of (cheaper) imports.

After 1971 other countries allowed the exchange rates of their currences to 'float'. Consequently MCAs were placed on a continuing basis, applied to all Members of the Community, and eventually credited to the budgets of the exporting country. The broadened scope of MCAs took account of the fluctuating divergence between national 'green' and market rates of exchange in the context of agricultural trade.[17] Moreover, in relation to national economic conditions, each government has been able to negotiate revaluations and devaluations of its own 'green' rate of exchange. Other developments occurred in the 1970s.[18] In 1973, for example, the AUA was revalued using an average of several currencies in what was termed the 'Joint Float'. MCAs for these currencies were calculated at a fixed rather than variable rate (monetary percentage) although Ireland, Italy and the United Kingdom opted out of the arrangement. Nevertheless, the 'Joint Float', together with the introduction of the European Monetary System in 1979, helped to bring about a marked decline in the size of MCA percentages by 1981 (Table 10).

The system of MCAs has maintained the principle of common prices as far as agricultural trade within the Community is concerned. But when 'green' rates have been used to convert common prices into national currency equivalents for domestic producers, variable price levels have been created (Table 8). In countries such as Italy, Ireland, and the United Kingdom, the 'green' rate has been maintained at a level above the market rate of exchange (Table 10). Producers have received lower prices than had the market rate of exchange been used to convert EC prices (AUA) into national currency equivalents. Conversely, in countries such as the Netherlands and West Germany where the 'green' rate has remained below the market rate, 'higher' agricultural prices have prevailed. In the autumn of 1976, for example, some agricultural prices in West Germany were nearly fifty per cent higher than in the United Kingdom. Thus in weak currency countries agriculture may have experienced a diminished ability to compete for national resources, whereas

Table 10. MCA percentage rates 1972–82 (January each year)[a]

Country	1972	1973[b]	1974	1975	1976	1977	1978	1979	1980[c]	1981[c]	1982[d]
W. Germany	+10.8	+2.5	+12.0	+12.0	+10.0	+9.3	+7.5	+10.8	+9.8	+6.5	+8.0
Belgium/ Luxembourg	+9.5	+1.3	+2.7	+2.7	+2.0	+1.4	+1.4	+3.3	+1.9	0	−8.1
Netherlands	+9.5	+1.3	+2.7	+2.7	+2.0	+1.4	+1.4	+3.3	+1.9	0	+4.0
Denmark	0	0	0	0	0	0	0	0	0	0	−1.8
Ireland	–	−7.0	−13.8	−10.5	−4.8	−23.5	−4.1	−2.0	0	0	0
France	+5.9	−2.5	0	−7.2	0	−17.5	−19.4	−10.6	−3.7	0	0
Italy	+4.9	−10.3	−13.0	−4.1	0	−19.2	−22.5	−17.7	−2.6	−1.0	−2.3
United Kingdom	–	−7.0	−13.8	−13.8	−6.4	−38.5	−31.6	−27.0	0	+12.9	+8.1

a: in simplified form:

$$\frac{market\ rate - green\ rate}{market\ rate} \times 100$$

b: March; c: April; d: February

Sources: R. W. Irivng and H. A. Fearn, Green money and the Common Agricultural Policy, Occasional paper 2 (Centre for European Agricultural Studies, Wye College, University of London, 1975); C. Ritson and S. Tangerman, 'The economics and politics of monetary compensatory amounts', European Review of Agricultural Economics, VI, (1979), pp. 119–64; Agra Europe (various years).

in strong currency countries the converse may have held. The probable aggregate result for the EC has been a misallocation of resources in agriculture[19] compared with the position determined by market exchange rates for national currencies. Moreover, since MCAs are paid only on those commodities for which there is an intervention system and considerable international trade,[20] the degree of misallocation seems likely to have varied on a product-by-product basis. The situation has been made more complex for those countries joining the EC in 1972. They have had to make an upward adjustment in their agricultural price levels so as to attain EC levels by the end of the five year transition period.

Conclusion

A number of significant ramifications flow from the methods of funding the CAP and setting agricultural price levels. The intervention system of the CAP, for example, shelters farmers from the discipline of overproducing for the market. Intervention agencies for designated products are required to purchase commodities that are offered to them and which meet the necessary quantity and quality standards. Producers are thus faced with an unlimited market at the intervention price. This system of support does nothing to discourage production that is surplus to market requirements and may indeed encourage such production. In 1978, for example, the costs of butter storage in the EC exceeded payments under the Regional Fund.[21]

In addition, some commodities are produced under closely regulated institutional prices and managed markets[22] – cereals, dairy products, beef and veal, sugar, and olive oil (Table 3). A second group experiences more modest levels of price support and protection from non-EC producers – pigmeat, eggs, poultrymeat, wine, and tobacco. The third group of agricultural products, characteristic of Mediterranean regions, receives relatively little support – fruit, vegetables, and minor arable crops. Individual farmers, therefore, have to operate under a variety of agricultural policy conditions. At one extreme they can be heavily dependent on institutional arrangements and suffer the political uncertainties of annually negotiated prices. At the other extreme the uncertainty of the free market can play the larger role in their production decisions. Any discussion of agriculture in the EC needs to take account of these distinctions in the way the CAP operates.

Turning to product price levels, the compromise package of decisions that is the outcome of the political bargaining in the annual price negotiations has tended to inflate the initial price proposals of the Commission. In most years, this has caused EC prices to diverge from lower world price levels. The magnitude of the difference represents an indirect, hidden cost of the CAP which can be estimated only with difficulty.[23] (Table 6). Moreover, the direct budgetary cost of supporting these prices has resulted in the Guarantee section of FEOGA increasingly dominating the Guidance section in the allocation of funds. It has also ensured that agriculture remains the most important claimant on the EC Budget bringing about net resource transfers between Member States. Unfortunately, the limits of the 'own resources' system of funding the Budget have been reached, and in addition the size of the transfers has become so large as to create conflict between national interests. Together these issues disrupt the smooth working and evolution of the Community and a revision of the method of funding the Budget, and hence FEOGA, is being sought.[24]

MCAs have both contributed to and alleviated pressures on FEOGA and the Budget. In one respect, they have become an additional and some would argue unnecessary expenditure from the Agriculture Fund.[25] On the other hand, Member governments have been able to negotiate changes in their own 'green rates' of exchange and so regain a degree of control over domestic agricultural price levels and net resource transfers. In 1977, for example, MCAs acted as a consumer subsidy on food imports to the United Kingdom to the value of £500 million, and for a time helped soften national attitudes towards funding the Budget. One important side-effect, however, has been the disruption of common institutional price levels within the EC. CAP prices have varied from country to country within the Community according to artificially maintained national 'green' rates of exchange. Consequently, agricultural production over the last decade has not developed under conditions of comparative economic advantage as initially envisaged for the EC.

Notes

1 This chapter draws heavily on R. Fennell, *The Common Agricultural Policy of the European Community* (London, Granada, 1979); and on M. Butterwick and E. N. Rolfe, *Agricultural Marketing and the EEC* (London, Hutchinson, 1971). Consideration is given to the price and marketing mechanisms of only seven major products in this book (cereals, dairy products, poultrymeat, eggs, pigmeat, beef/veal and sugar). The reader should consult the above texts for regulations applying to other agricultural products.

2 D. Coombes, *The power of the purse in the European Communities*, European Series 20, PEP (London, Chatham House, 1972).

3 M. Whitby, *The net cost and benefit of EEC membership*, Seminar Paper 7 (Centre for European Agricultural Studies, Wye College, University of London, 1979).
4 Although introduced in 1975, the European Unit of Account (EUA) has been used in financial calculations for the Budget only since 1978. In former years the Budget was expressed in the Unit of Account (UA). The EUA is a 'basket' of Member States' currencies weighted according to each country's share of the Community's GDP, intra-European trade and the currency aid mechanism. Its value is calculated from day to day for each currency on the basis of the market exchange rates at the close of trading. The EUA was renamed the European Currency Unit (ECU) in 1979 for use in the new European Monetary System (EMS). European Documentation, *European Economic and Monetary Union* (Brussels, Office for Official Publications of the European Communities, 1981).
5 J. De Veer, 'The objective method: an element in the process of fixing guide prices within the CAP', *European Review of Agricultural Economics*, VI, (1979), pp. 279–302.
6 Fennell, *The Common Agricultural Policy*, pp. 55–9.
7 Pilot group number, 1: whey powder, 2: whole milk powder, 3: skim milk powder (SMP), 4: unsweetened condensed milk, 5: sweetened condensed milk, 6: Blue-veined cheese, 7: Parmesan-type cheese, 8: Emmental, 9: Gouda, 10: Butterkase (butter 82% fat), 11: Cheddar, 12: Lactose.
8 Producers contribute towards the cost of disposing of surplus dairy products by a levy on milk delivered to a dairy or processed on the farm. Initially the levy was set at 1.5 per cent of the target price for milk. By 1978/9 it had been reduced to 0.5 per cent, but had risen in stages to 2.5 per cent by 1981/2. Producers in defined mountain and hill areas are excluded from the full levy.
9 This measure (Reg. 1078/77) assists dairy farmers to give up milk production or convert their dairy herds to beef production. It continued the Dairy Herd Conversion Scheme of October 1969 (Reg. 1975/69) as extended to larger herds in May 1973 (Reg. 1353/73). The producer must agree not to dispose of milk either by sale or free of charge for a period of at least five years.
10 Weighting is proportional to the size of the cattle population in each country so that depressed markets in one country do not have a disproportionate effect on the final Community reference price.
11 For any Member State, intervention can also be activated when the national reference price for mature cattle falls below 98 per cent of the guide price and remains below the intervention price of two consecutive weeks. Conversely, intervention is suspended when the reference price has been greater than the intervention price for three consecutive weeks. In practice, these arrangements are more complex as they vary in detail with different categories of beef product.
12 Subject to the constraint that in any year the value of the premium, plus the price realised for fat cattle, must not exceed 85 per cent of the guide price on average, or 88 per cent of the guide price at any specific point of time.

13 Fennell, *The Common Agricultural Policy*, p. 129.
14 T. Heidhues *et al*, *Common prices and Europe's farm policy*, Thames Essays 14, (London, Trade Policy Research Centre, 1978), p. 14.
15 A full treatment is given by R. W. Irving and H. A. Fearn, *Green money and the Common Agricultural Policy*, Occasional Paper 2 (Centre for European Agricultural Studies, Wye College, University of London, 1975).
16 Fennell, *The Common Agricultural Policy*, p. 91.
17 A calculation is set out in C. R. Groves, *An EEC agricultural handbook*, Research and Development Publication 6 (Ayr, West of Scotland Agricultural College, Auchincruive, 1978), p. 45–6.
18 The generalised system of MCAs was established under Regulation 974/71 and was subsequently amended by Regulations 222/73 and 1123/73. In 1973 a 'representative' rate for each country replaced the parity rate declared to the International Monetary Fund for converting AUA into national currency equivalents.
19 Irving and Fearn, *Green money and the Common Agricultural Policy*, p. 38.
20 MCAs on poultry and eggs are an exception since these products are based on cereals which have intervention mechanisms. Two additional complexities have been added to the MCA system: MCA percentages are allowed to vary between products within each Member State; changes in 'green' rates take effect at different times for each commodity sector following the annual price negotiations.
21 J. Cooney, *A United State of Europe?* (Dublin, Dublin University Press, 1980), p. 67.
22 Fennell, *The Common Agricultural Policy*, p. 111.
23 U. Koester, 'The redistributional effects of the common agricultural financial system', *European Review of Agricultural Economics*, V, (1978), pp. 321–45; A. E. Buckwell, *et al*, *The costs of the Common Agricultural Policy*, (London, Croom Helm, 1982), pp. 58–9.
24 J. Pearce, *The Common Agricultural Policy: prospects for change* (London, Chatham House Papers, 1981).
25 Academic opinion is divided: see C. Ritson and S. Tangerman, 'The economics and politics of monetary compensatory amounts', *European Review of Agricultural Economics*, VI, (1979), pp. 119–64.

6

The national concentration of agricultural production

The creation and subsequent evolution of the CAP coincided with a period of major developments in agricultural technology within Western Europe. Progress in crop science, for example, resulted in improved varieties of crops, new chemical fertilisers and crop-protection sprays, more powerful and effective machinery to enhance the timeliness and speed of farm work, and improved harvesting equipment and storage facilities.[1] With the application of these developments to farming throughout Western Europe, purchased inputs to agriculture have increased through time leading to the intensification of output per hectare and the modernisation of production techniques. In France, for instance, between the mid-1950's and 1980/1, the use of fertilisers rose from 1.5 million to 5.6 million tonnes per annum, the number of tractors increased from 337,000 to 1.5 million, and the yields of wheat and barley expanded from 1.8 to 4.8 and 1.6 to 4.2 tonnes per hectare respectively.[2]

However the CAP cannot take credit for the intensification and modernisation of agricultural production. Not only are these developments observable in other developed market economies around the world, but they were emerging before the CAP was made effective in each Member State. On the other hand, the CAP has undoubtedly altered the economic environment within which farm modernisation has taken place. Here recent agricultural trends in the EC are examined with a view to evaluating the effects of the CAP and in particular the impact on the location of agricultural production.

EC production trends

Changes in the aggregate structure of EC agricultural production are displayed in Table 11 using 'current value' as the measure.[3] Over the past two decades the broad balance of EC agriculture has remained relatively unaltered. Meat and livestock products continue to dominate accounting for nearly sixty per cent of total agricultural output, while in recent years the relative value of all crop production has fallen slightly to nearly 40 per cent.[4] The role of milk has been consolidated as the single most important product in EC agriculture (19 per cent of total output); significant but lower proportions of output are accounted for by beef/veal (15 per cent) and pigmeat (bacon, ham, pork – 12 per cent). Cereals in aggregate also account for 12 per cent of output, with wheat remaining the most important crop. Other products, such as wine, fresh fruit, and eggs, contribute a small and generally stable or declining proportion to the total value of output. Only barley has shown a marked increase in the share of production in recent years; this reflects the surge in EC grain production in the 1960s and 1970s when the ratio of crop prices to the cost of fertiliser was particularly favourable.[5]

The intensification of agricultural production in the EC (The Nine) is described in Figure 6a for the period 1969–81. Three groups of products can be identified: those with high rates of increase; intermediate increases; and low increases or decreases in production. However, the small initial (1969) volume of some commodities such as maize grain, and conversely the large

Table 11. The structure of agricultural production in the EC (% total current value)

Product	1963 (The Six)	1974 (The Nine)	1982 (The Nine)	1982 (The Ten)
Milk	18.3	17.7	19.7	19.2
Beef & veal	17.1	16.2	15.1	14.6
Pigmeat	12.8	11.8	12.3	11.9
Wheat	7.1	6.9	7.2	7.3
Fresh vegetables	7.9	7.2	6.5	6.7
Wine	5.7	6.6	6.0	5.8
Poultrymeat	4.3	4.3	4.4	4.4
Fresh fruit	5.2	4.2	3.7	3.9
Eggs	5.4	4.2	3.0	3.0
Barley	1.7	2.0	3.3	3.2
Sugar (beet)	2.3	2.4	2.4	2.4
Potatoes	2.6	2.1	2.2	2.2
Sheepmeat	n.d.	1.1	1.4	1.8
Maize grain	0.9	1.7	1.8	1.9
Olive oil	2.1	1.5	0.7	1.2
Industrial crops	1.2	1.3	0.9	0.9

Source: European Communities, *Agricultural Statistics* 3 (Brussels, 1975), pp. 28–9. Commission of the European Communities, *The Agricultural Situation in the Community, 1983 Report* (Brussels, 1984), pp. 190–1

volume of others such as milk should be borne in mind when interpreting the diagram. The impact of bio-technical progress in agriculture is evident in the increasing output of most commodities.

Maize grain, wheat, poultrymeat, and pigmeat exhibit above-average rates of increase and bear a relationship to each other. Both poultry and pigs have been farmed increasingly under capital-intensive, factory-like systems of production in which livestock convert feed grains such as maize into a meat product. Housed in large numbers in closely regulated environmental conditions, the livestock are produced in high volume and at reduced unit costs of production because of scale economies. For example, between 1973 and 1979 the number of holdings with pigs in the EC fell by 4.5 per cent, especially those with fewer than 200 pigs. The average herd size rose over the same period from twenty-five to thirty-five animals and to forty pigs by 1981. Similar changes are evident for poultry production. By 1977, 73 per cent of broiler production was concentrated on 0.5 per cent of the holdings with broilers and in flock sizes in excess of 10,000 birds.

The expansion of poultrymeat and pigmeat production has been facilitated by an increasing per capita consumption within the EC (Table 14), especially in West Germany and Denmark for pigmeat, and in Italy and France for poultrymeat. Poultrymeat has also received a stimulus from a strong growth in exports to the Middle Eastern countries of Saudi Arabia and Iran. Exports to these countries rose from 16,157 million tons in 1973 to 110,264 million tons in 1977 alone. France, and more recently the United Kingdom and West Germany have benefited from this trade which has recently attracted increased export subsidies as demand in the Community has become saturated.

The expansion of poultrymeat and pigmeat production has increased the demand for and stimulated the supply of feed grains (Table 12), although rising yields rather than the increasing area under cereals, especially of wheat, appears to have contributed most to the growth of production.[6] Grain maize is an exception, however, for the development of hybrid varieties suited to the shorter growing season of northern parts of the EC has allowed the area under the crop to be expanded. A rising demand for feed-grains has also come from milk, beef, and veal producers. Their systems of production have been intensified by increasing the numbers of livestock per hectare of agricultural land, with a consequent greater dependence on purchased feed. The CAP has contributed to the expanding volume of cereal production by maintaining target and intervention prices

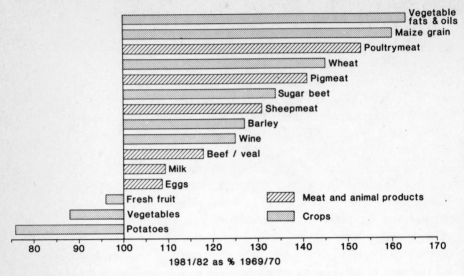

Figure 6A. EC production changes, 1969/70–1981/2 (volume)

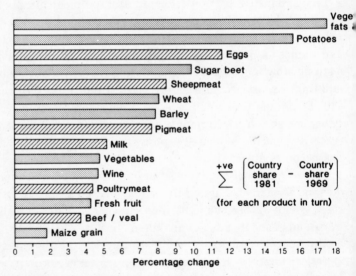

Figure 6B. Changing national concentrations of production, 1969–81 (volume)

above those ruling in the world market, and by providing an assured market through intervention when production has been in excess of demand inside the Community.

One unfortunate side effect of these developments in relation to the CAP

Table 12. Products used for animal feeding in the EC (% total)

Product	1972/3	1980/81	% change 1972/80 (volume)
Cereals (feed grains)	78.0[a]	70.5	−0.4
Feed cake	15.8 (8.4[b])	23.1 (14.6[b])	+5.2 (+7.6[b])
Animal meal	2.4	2.4	+2.9
Dehydrated fodder	1.5	1.9	+0.9
Skim milk powder (SMP)	1.3	1.6	+2.3
Legumes	0.9	0.6	−1.5
Total ('000t)	86,596	97,502	+12.6

a: of which 82% produced in EC; b: soya
Source: Commission of the European Communities, *The Agricultural Situation in the Community, 1983 Report*, (Brussels, 1984), p. 418

emanates from the use of cereal substitutes. Feed manufacturers and livestock farmers have sought out substitutes for the relatively expensive cereals produced in the EC. Soya beans and maize gluten have been imported in increasing quantities from the USA, manioc from Thailand, cereal residues from Argentina, and molasses from Brazil and Cuba. Consequently, the EC has become the world's largest importer of products for animal feed, a position that is favoured by the absence of CAP levies on such imports. Indeed the production of animal feedstuffs has become so sensitive to variations in world market prices that highly sophisticated mathematical techniques of linear programming are used for optimising the input from various sources to the animal rations.[7] The effect of cereal substitutes has been to reduce the input of EC cereals into livestock feedstuffs (Tables 12 and 13) so that wheat in particular has become an insignificant (12 per cent) proportion of the final product. Wheat production is also faced by a declining per capita consumption through the medium of bread, cakes and biscuits (Table 14).

The rising volume of cereals produced in the Community, therefore, now exceeds internal demand and expenditure on export restitutions has increased as a result of dispersing the surplus grain on the world market. The rather absurd situation exists whereby livestock production in the EC has become increasingly dependent on imports of cereal substitutes (protein) from developing countries (6.35 million tonnes in 1979), while the export of cereals is subsidised. A corrective reduction in CAP cereal prices to world levels is now being attempted in stages, but proposals to impose levies on imported feed proteins have so far been resisted. Exporters such as the USA, for example, have protested that their markets would be restricted, consumer interests have opposed the higher meat prices that would follow any increase in animal feed costs, while meat producers in strong-currency countries, such as Belgium, West Germany and the Netherlands, have supported the arrangements since at present they receive a cost advantage in purchasing feed.

Turning to vegetable fats and oils (sunflower, colza, rapeseed, castor seed, linseed, cotton seed), the volume of production has nearly doubled since the mid-1960s (Figure 6a). Initially vegetable oils suffered technical problems: the low quality of sunflower seeds, for example, and poor nutritional quality of the first varieties of colza through high levels of erucic acid. As improved varieties overcame these problems, and the price ratio with cereals improved, so the production of oilseeds increased. A rising demand from the food industries producing margarine, cooking oils and cattle feed also gave bouyancy to the crops, and proposals to increase the import levy on vegetable fats (oilseeds) to make them less competitive with dairy products in the manufacture of margarine have so far been resisted. At present oilseed rape is associated with northern areas of cereal production and sunflower and soya with southern regions of the EC. In many parts of the Community, however, the ceiling on the area under oilseed crops is being reached and further expansion seems dependent on the development of new varieties to combat present climatic limitations.

Sugar, in the form of sugar beet, is another product that has greatly increased in volume under the CAP despite the operation of production quotas. Indeed, in most years since the mid-1970s the Community has

Table 13. Utilisation of raw materials in animal feedstuffs – Netherlands (%)

Year	Cereals	Oilseed cakes and meals	Other products (of which manioc)
1960/1	63.0	13.0	24.0 (0)
1963/4	59.8	14.8	25.4 (0)
1966/7	47.4	18.9	33.7 (1.5)
1970/1	33.7	22.9	43.4 (5.6)
1974/5	31.3	24.3	44.4 (10.7)
1978/9	16.6[a]	26.5	56.9 (16.9)

a: EC average 39%
Source: M. L. Debatisse, *E.E.C. organisation of the cereals markets*, Occasional Paper 10, (C.E.A.S., Wye College, 1981), p. 29

produced 20 per cent more sugar than it consumes. In 1979, for example, 1.4 million tons out of a total of 12.3 million tons was produced as C category sugar. The problem can be traced back to 1974 when sugar quotas and prices were raised at a time of shortages and high prices on world markets. With hindsight decision makers over-reacted to a temporary world shortage of sugar[9] but the result was a rapid and substantial expansion of production by 1.8m tonnes. Concern over possible world sugar deficits, together with national disputes over the allocation of production quotas, have thwarted subsequent attempts to limit production. The problem has been made more severe by a falling per capita consumption of sugar in the EC since 1973.

A second group of products exhibit more modest rates of increase in output and include sheepmeat and eggs. Per capita consumption of these products has been static or falling (Table 14) and both have received relatively little price support under the CAP. Sheepmeat benefited from market regulation after 1980 but the effect on production is not measured by the available statistics. In the same group are the major commodities of beef and milk, both suffering from falling levels of per capita consumption. Increased production in both cases has come from the use of compound animal feeds, disease control and breed selection through artificial insemination. Milk production, for example, has been increasing at an average annual rate of 1.7 per cent since the early 1960s; estimates suggest that as much as 20 per cent of this output has been generated by compound feed since approximately 60 per cent of total oilcake production is consumed by dairy cows. Certainly, the average yield of dairy cows in the Community has continued to rise (1.4 per cent a year) despite a fall in the number of producers and a gradual decline in the number of dairy cows (0.2 per cent a year).[10] In addition the dairy sector has received price increases over the last decade which have done little to correct the severe problems of oversupply of milk and dairy products.

Wine is also part of the group or products having modest rates of increase in production under the CAP. The output of wine fluctuates markedly from year to year, according to weather conditions, but has displayed a long-term increase of 0.56 per cent a year since 1970.

The third group of products with relatively static or declining output is exemplified by vegetables, fresh fruit and potatoes. Per capita consumption of all three products is stable or falling in the EC (Table 14), and they are also without well-developed CAP market regulations. Nevertheless, there are marked national differences in consumption patterns. The demand for vegetables is higher in France and Italy than elsewhere in the Community, whereas the consumption of fresh fruit is higher in West Germany and the

Table 14. EC levels of self-sufficiency (%) and per capita consumption (kg/head)^a

Product	Self-sufficiency^b			Consumption^b		
	1967–9 (ave.)	1975–7 (ave.)	1979–80^c	1967–9 (ave.)	1975–7 (ave.)	1979–80^c
Butter	91	111	119	6	6	5
Sugar	82	111	124	36	36	36
Poultrymeat	101	104	105	9	13	14
Barley	103	103	112	–	–	–
Wheat	94	100	117	76	74	77
Pigmeat	100	100	100	28	34	37
Wine	97	98	105	51	49	47
Potatoes	100	98	101	90	73	77
Beef	90	97	98	25	25	22
Total Vegs.	98	93	98	99	99	107
Fresh fruit	80	77	83	65	58	62
Rice	n.d.	64	83	n.d.	3	3
Sheepmeat	56	66	71	3	3	3
Grain maize	45	50	62	n.d.	3	3

a: changed definitions make it impossible to calculate consistent data for the 1950s; b: The Nine; c: The Ten
Source: Commission of the European communities, The Agricultural Situation in the Community. 1982 Report (Brussels, 1983), pp. 242–5. European Community, Green Europe Newsletter 13 (1980), Table 5

Netherlands. Potatoes are more popular in Ireland, Belgium, and the United Kingdom but only small quantities are consumed in Italy.

National production trends
Production trends under the CAP have varied from country to country for each commodity. They can be described by calculating the national shares of each product for two time periods, in this case 1969 and 1981 (Table 15). Comparable data for The Nine cannot be assembled until 1969, and prior national trends in The Six are not considered in full here.[11] To aid interpretation of the data, the reader should bear in mind the national shares of the agricultural resource base of the EC. The total utilised agricultural area is a convenient measure although there are variations in definition from country to country.[12] For example, the United Kingdom statistics include large areas of very poor grazing and almost unutilised tracts such as deer forests in Scotland; in other countries these areas would be classified as 'other land'. The size of the work force, the number of farm units, the total capital investment or even the total arable area could be advanced as alternative

measures of the resource base, but they contain even more severe problems of definition. The degree of 'concentration' of production can be judged by dividing each national share of a product by the national share of the agricultural resource base shown in Table 15. Known as the 'location quotient', the figures above unity signify a share in excess of the resource base (concentration), while figures below unity indicate the relative absence of concentration of the product in that country. In Table 15, however, only the raw percentage data are presented.

France and Italy, either individually or jointly, dominate the production of seven of the eight most concentrated products in the EC. Maize grain is the most concentrated of the fifteen commodities under consideration, 60 per cent of production (by value) emanating from France and a further 30 per cent from Italy. Also located in these two countries are 80 per cent of wine and 60 per cent of wheat production, while France alone accounts for 52 per cent of the Community's output of oilseeds. Fresh fruit and fresh vegetables are concentrated in Italy.[13] For most of these highly concentrated products, the share of either France or Italy has been increasing rather than decreasing under the CAP. There are detailed variations, however, within these broad product labels. Under the heading of vegetables and fruit, for example, tomatoes have become more concentrated in France and the Netherlands at the expense of Italy, apples have become more concentrated in France compared with all other countries, while cauliflowers have increased in concentration in the United Kingdom and France but have declined in importance in Italy and Belgium.

Of course, countries other than France and Italy have increased their share of these highly concentrated products but in a less systematic way. West Germany, for example, now produces a greater share of oilseeds, wheat and wine in the EC compared with a decade ago, but has a smaller share of maize grain, fresh fruit, and vegetable production. In more detail, the country's wine sector has been substantially restructured in recent years.[14] Almost half of the previously fragmented vineyards have been consolidated, resulting in reduced labour costs and increased mechanisation. Yields have been increased by planting grafted wine stocks of new grape varieties which are frost resistant and productive under the cool, moist conditions of northern Europe. Reisling varieties are still the top quality but Morio-Muskat, Scheurebe and Kerner grapes have increased in area and production. This expansion has been facilitated by a rising domestic consumption in West Germany – from 15.7 litres per head in 1969 to 25 litres in 1979 – and by increased exports.

Table 15. National concentration of agricultural production, 1969–81 (% total value of EC output)

Country	Utilised agricultural area	Maize grain	Wine	Oilseeds	Fresh fruit	Wheat	Barley	Fresh Veg.
West Germany	12	3	16*	16*	14	12*	17*	5
France	31	60*	48*	52	22*	39*	27	21
Italy	18	30	32	4	40*	21	1*	39*
Netherlands	2	0	0.2*	2	3*	2	1	10*
Belgium/Luxembourg	1.6	0	0	0	3*	2	2	5
United Kingdom	18	0	0	16*	6	17*	33	12
Ireland	6	0	0	0	0.2*	0.4	4*	1*
Denmark	3	0	0	10*	0.6	2*	13	0.8
(Greece)	(9)	(7)	(3)	(0.5)	(11)	(5)	(1.4)	(7)
PBCCa	—	+.19	+.01	-.20	+7.3	+.30	-.35	+.06

Country	Poultry meat	Sheep meat	Sugar (beet)	Beef and veal	Milk	Pigmeat	Potatoes	Eggs
West Germany	7*	3*	25*	20*	23	31	17	20
France	31	28*	27*	29	24	16*	14	20
Italy	29*	9	17	15	12	11*	15	16*
Netherlands	8	3*	7	7*	12*	12*	13*	10*
Belgium/Luxembourg	2	0.4*	6	4	3	7*	5*	4
United Kingdom	16	29	9	15	16*	10	26*	21
Ireland	1	4	2*	5*	4*	1	1*	2
Denmark	2	0	3	3	5*	10	2*	2
(Greece)	(4)	(24)	(3)	(1.6)	(2.1)	(1.9)	(7)	(5)
PBCCa	+.19	-.29	+.40	-.11	-.48	-.23	-.15	+.11

a: Point Biserial Correlation Coefficient between trend and degree of concentration in 1969 (excluding Greece), + = more concentrated, − = more dispersed distribution; *: increasing share 1969–81
Source: author's calculations from agricultural statistics

The United Kingdom, like West Germany, has also increased its share of wheat production and oilseeds. Neither crop can be said to be concentrated in the United Kingdom, however, despite an increase in the production of wheat from 4.24 million tonnes to 7.14 million tonnes between 1970 and 1980 which has helped to turn the country from a net importer to a net exporter of cereals.

Barley is the only highly concentrated product not associated with France or Italy. The crop is concentrated in the United Kingdom, West Germany, and Denmark, although yields are lower in Denmark than elsewhere. When taken in aggregate cereals have become more concentrated in France and West Germany at the expense of the Netherlands. Indeed, France has shown the most rapid increase in cereal yields, especially wheat, since 1957, and yields in most Member States exhibit a linear trend indicating the prospect of further increases in production in the years ahead.[15]

The remaining seven commodities described in Table 15 have lower and more variable patterns of national concentration in the EC. Sheepmeat, for example, is concentrated in the United Kingdom, but the product is becoming increasingly important in France, West Germany and the Netherlands. Sugar beet is becoming even more concentrated in West Germany and France at the expense of all other Member States except Ireland. The long-term (1957–78) trend of rising yields has been highest in France and Ireland, although there is evidence of a levelling off in the rate of increase of yields throughout the EC.[16] To an extent, these adjustments in the concentration of sugar beet were established by the allocation of production quotas in 1968 which favoured areas best suited to production. At that time, Italy suffered a 25 per cent reduction in the area under sugar beet while France gained by 35 per cent.[17] In the United Kingdom, by comparison, yields have not increased greatly since 1965, and because of the profitability of cereals the national basic quota for sugar production has not been fulfilled on occasions in recent years. Changes in the location of production are gradual, however, since the imposition of quotas at the outset was designed to prevent rapid changes in the national pattern of output.[18]

Even crops in decline, such as potatoes, show a changing pattern of national concentration under the CAP. The Netherlands has increased its share of the declining production by mechanising farming practices especially on the clay lands in the north and west of the country. In addition, new potato varieties, an increase in the industrial processing of the crop (chips, crisps etc.) and rising exports both to other Member States and countries outside the EC have served to maintain the profitability of the crop. There have also been changes in the structure of potato production.[19]

Given the need to operate machinery more effectively on large crop areas, the average area of potatoes per farm rose from 2.2 to 5.7 hectares between 1965 and 1978, while the proportion of producers with more than 10 hectares of potatoes increased from 25 to 51 per cent.

Attention is now turned to three different summary measures of these complex trends in the national concentration of production under the CAP. First, changing national shares of each product are summarised in Figure 6b for the period 1969 to 1981. Larger values indicate greater changes in the distribution of a commodity within the Community, while small values show a relatively constant distribution of the crop or livestock product. Table 15 indicates if the overall result has been one of greater concentration or dispersion amongst all the Members of the Community.[20] Eight products exhibit significant changes in their distribution amongst the Member States: eggs, sugar beet and wheat becoming more concentrated, and vegetable fats and oils, potatoes, sheepmeat, barley, and pigmeat becoming more dispersed. These can be termed 'redistributing' products although with the exception of pigmeat and potatoes they showed lower rates of change in their distribution under the Community of The Six. It should be noted that the products have no similarities in aggregate production trends (Figure 6a) or types of CAP market regulation. Products both with (wheat, pigmeat) and without (eggs, potatoes) closely regulated institutional prices and managed markets are in the same group.

The remaining seven products have been more stable in their distribution: vegetables, wine, poultrymeat, fresh fruit and maize grain have become slightly more concentrated within the EC as a whole, while milk and beef/veal are now marginally more dispersed than a decade ago. Only poultrymeat and vegetables showed high rates of 'redistribution' in the community of The Six. Once again, there is no commonality in the degree of market regulation under the CAP for these production trends.

A second summary measure has been created by grouping together those countries which have experienced broadly similar trends (shares) in production between 1969 and 1981. Two groups can be discerned (Table 16), although it must be emphasised that there is some variability in trends within each set of countries. Group I comprises the three countries which joined the Community in 1972 (Denmark, Ireland, the United Kingdom) together with West Germany. This group has tended to enjoy rising shares of total EC output of wheat, potatoes, milk and oilseed, but a falling concentration of maize grain, fresh fruit, pigmeat, eggs and sheepmeat. The remaining countries comprise Group II and in general exhibit converse trends to Group I. France and Belgium/Luxembourg have the closest similarity of trends including increasing shares of pigmeat, fresh fruit and sheepmeat, but declining proportions of barley, oilseeds and beef/veal. Interestingly, West Germany also experienced different production trends from other countries in the Community of The Six, while the Netherlands and Belgium/Luxembourg had the greatest similarity in agricultural changes at that time.

Table 16. National trends in the concentration of agricultural production, 1969–81

	Country	Increasing shares	Decreasing shares
Group I	Denmark	wheat	maize grain
	United Kingdom	potatoes	fresh fruit
	Ireland	milk	pigmeat
	West Germany	oilseeds	eggs
			sheepmeat
Group II	France	fresh fruit	wheat
	Belgium/Luxembourg	pigmeat	barley
	Netherlands	eggs	oilseeds
	Italy	sheepmeat	sugar beet
			beef/veal
			poultrymeat
			milk

Source: author's calculations from agricultural statistics

A third facet of the concentration of production is revealed by the changing total volume of agricultural output. For the period since 1963, national shares of total EC production have shifted in favour of the Netherlands, Ireland and Belgium/Luxembourg at the expense of Denmark, the United Kingdom and West Germany.[21] Nevertheless, there have been fluctuations within these broad trends: since 1970, for example, Denmark and West Germany have experienced more rapid rates of increase in total output than previously, whereas the rate of increase has fallen in France; the rate of growth of production in the Netherlands was at its highest prior to 1973.[22]

These national figures disguise marked regional variations within each Member State in the concentration of total production (Figure 7b). Rates of increase in output per hectare have been greatest in southern and central Ireland, northern Italy, northern and southern West Germany,[23] the

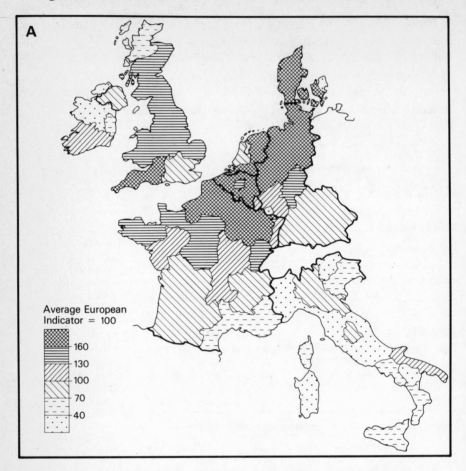

Figure 7A. Estimated FEOGA expenditure by region per person employed in agriculture, 1976/7

Netherlands, Belgium, Britanny and Côte d'Azur. At the other extreme, growth rates have been very low throughout central and south-west France, Denmark, parts of central and southern Italy, Sicily and Sardinia.

Factor costs and product prices
Attention so far has been concentrated on a description of recent trends in agricultural production in the EC. No consistent relationships have been found between those trends and the various types of market regulation under the CAP. Rather the roles of changing demand (per capita consumption),

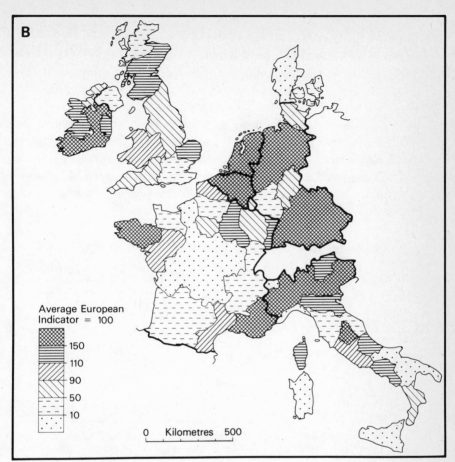

Figure 7B. Average annual growth of production per hectare, 1968/9–1976/7

new agricultural technology and modernised systems of farming have been stressed. These factors alone, however, are insufficient in explaining the production trends, especially when they vary so markedly from country to country. Consequently, consideration is now given more fully to the broad economic environment within which each national agriculture has to operate. A division is made between factor (input) costs and product (output) prices, with the CAP exerting an important influence over the latter.

Factor costs

In a common market, internal barriers to trade are removed and a uniform method of deriving product prices operates. Under such conditions it is expected that different countries inside the market would gain a comparative advantage in the production of certain commodities based on lower unit costs of production, marketing and transport to the point of consumption. In the context of the EC, variations in soil and climate are expected to give Member States such as France and the United Kingdom cost advantages in the production of cereal crops; Mediterranean countries of the EC should enjoy lower costs in the production of fruit, vegetables and wine; countries near to the North Sea ports ought to benefit from cost advantages in access to imported animal feed; the moister northern and north-western Members of the Community should have advantages in the production of grass for livestock-rearing enterprises.

Table 17. Changing national prices of factor inputs, 1970–6 (1970 = 100)

Country	Machinery	Buildings	Feedstuffs	Fertiliser
Ireland	284	266	263	301
Italy	235	254	241	215
United Kingdom	252	243	263	204
Denmark	190	190	197	191
France	184	185	168	192
Belgium	160	206	147	201
Netherlands	166	176	127	178
West Germany	150	144	133	148

Source: R. Behrens and H. Haen, 'Aggregate factor input and productivity in agriculture', *European Review of Agricultural Economics* VII, (1980), pp. 109–46

Of course lack of 'natural' cost advantages can be overcome to an extent by purchasing factor inputs such as fertilisers, machinery and heating fuel, or by obtaining economies of scale through the development of a structure of large farm units and the shedding of farm labour. For example, the large farm structure of the United Kingdom has been estimated to give an 11 per cent cost advantage over agriculture in West Germany.[24] Equally, 'natural' advantage can be further enhanced by the purchase of these 'non-farm' capital inputs and by changes in farm structure. Unfortunely, the pricing of factor inputs has not been equalised within the EC, nor have changes in factor costs been uniform in the last two decades. The impact of price inflation has varied from country to country according to the economic performance of each national economy (Table 8). Between 1970 and 1980, for example, Ireland, Italy and the United Kingdom experienced the highest national rates of price inflation in the Community. In addition, inflation has effected different factor inputs to a varied extent (Table 17). For example, the price of fertilisers and machinery has been most adversely affected in Ireland, whereas in Italy the cost of acquiring buildings and animal feed has increased by the greatest amounts. By comparison, the rate of price inflation has been low in West Germany, Belgium and the Netherlands, especially for animal feeds, although increased interest rates in the 1970s raised the cost of production for those who had borrowed capital for investment.

Further, national governments have influenced the price of factor inputs. Dutch glasshouse producers, for example, have enjoyed subsidised prices for heating fuels, while in France horticultural producers have been eligible for a 30 to 35 per cent subsidy on the cost of capital investments – in the United Kingdom similar investments attracted only a 15 per cent national subsidy. State aids to farming remain entrenched and extensive features of agriculture in the EC, one estimate for 1977 valuing them at 13,615.3 million ECU in total. National aids in France, West Germany and the United Kingdom amounted to 5,703 million, 2974m. and 2188m ECU respectively.[25]

The ratio between the costs of acquiring inputs and the prices of farm products has influenced the way in which national agricultures have developed under the CAP. The input of fertilisers, for example, has increased by the greatest amount in West Germany in recent years (Table 18). The low national rate of increase in the price of fertilisers resulted in a high ratio of cereal prices to fertiliser costs thus raising the optimum level of application of fertilisers per hectare.[26] Fertilisers have also been substituted for increasingly expensive land by farmers wishing to expand their production of crops. In contrast, the price of fertilisers has increased by the greatest amount in Italy and Ireland, and these two countries also exhibit the lowest rate of increase in the input of fertiliser per hectare.

Changes in other inputs have also varied nationally.[27] Increases in the input of feeding stuffs have been greatest in France, Belgium, Italy and the Netherlands; machinery inputs have expanded at the highest rate in France, the Netherlands and Italy; investment in buildings has shown the greatest increase in Ireland and the United Kingdom. In some cases, the growth in factor inputs has not been in the national interest. There is some evidence to suggest that agriculture in the United Kingdom, for example, has become over-capitalised through the increased investment in farm buildings;[28] the national efficiency of agriculture could have suffered as a result. Interestingly, total inputs to agriculture have fallen in Denmark and West

Table 18. Fertiliser input per hectare of agriculturally used land (EUA/hectare)

Country	1963	1970	1977	Increase 1963–77
West Germany	37.4	49.9	58.4	+21.0
Netherlands	39.6	54.0	57.8	+18.2
France	14.3	24.1	31.9	+17.6
Denmark	25.7	35.0	41.4	+15.7
Belgium/Luxembourg	–	–	52.3	–
United Kingdom	16.3	22.2	26.4	+10.1
Ireland	7.7	11.5	15.7	+ 8.0
Italy	10.8	13.7	17.3	+ 6.5
(EC)	18.4	26.5	32.2	+13.8

Source: as Table 17

Germany because of substantial reductions in the labour force (wages), while total inputs have risen to the greatest extent in the Netherlands, Ireland and Belgium/Luxembourg.

Thus farmers in different countries have operated under very artificial conditions of comparative advantage. 'Natural' cost advantages or disadvantages have been distorted by the changing costs of factor inputs, changes that owe more to the varying performance of national economies in general than to the agricultural sector itself. Moreover, since the impact of price inflation has varied with the type of factor input, and different sectors of agriculture use each factor input in varying amounts, so the comparative advantage of producing each product has varied between the Member States. A recent study of the dairy sector, for example, found that between 1974 and 1980 production costs had risen to the greatest extent in Italy, France and the United Kingdom, and by the least amounts in West Germany, Denmark and the Netherlands, with consequences for the competitive advantage of each national dairy sector.[29]

Product prices

Producers have not been faced by uniform changes in product prices under the CAP (Tables 8 and 19). The prices of some commodities have increased at a faster rate than others: favourable price increases have been allowed for hard (durum) wheat, olive oil and citrus fruit, for example, reflecting the recent concern of the Community with the low incomes of producers in the Mediterranean regions of the EC. Oilseed rape, milk and barley, by comparison, have encountered relatively modest price increases in the 1970s. Thus price ratios have moved in favour of maize at the expense of wheat and barley, to the advantage of beef compared with milk and, until the late 1970s,

Table 19. Changes in producer prices of agricultural products, 1967–80 (1972/3 = 100)[a]

Product	1967/8	1971/2	1975/6	1980/1
Hard wheat	92.6	96.2	205.1	382.2
Olive oil	92.4	95.2	203.4	368.4
Citrus fruit (ave.)	94.6	93.7	180.9	358.5
Table wine (ave.)	–	93.3	154.7	275.7
Maize grain	82.4	95.2	137.6	239.6
Beef and veal	–	96.0	156.2	238.3
Common (soft) wheat	89.8	96.2	133.5	222.7
Tobacco (ave.)	–	96.6	137.3	222.1
Sugar beet	–	96.3	140.9	206.6
Milk	–	92.3	139.9	206.3
Barley	84.8	96.1	125.8	198.1
Pigmeat	90.7	97.0	134.5	192.5
Oilseed rape	92.3	97.1	123.3	181.5
Average all prices	90.9	95.4	145.2	225.8

a: expressed in national currency excluding VAT
Source: European Community, *Green Europe Newsletter* 13 (1980), Table 4.3

in favour of pigmeat in relation to barley. But these aggregate price changes bear no close relationship either to aggregate or national production trends in the EC, in part because variations in production (factor) costs are also involved in farmers' decisions on what to produce, and in part because product prices have varied markedly from country to country. Two sources of variation in product prices can be discerned.

First, national agricultures have experienced varying price increases as countries have made the transition to common institutional prices under the

Table 20. National product prices, 1966/7 and CAP prices, 1967/8 (UA/100 kg)

Country	Wheat	Barley	Milk
Italy	11.73	8.60	10.75
West Germany	10.54	10.57	10.05
Netherlands	10.20	9.01	9.55
Belgium	9.28	8.28	9.78
France	8.49	7.68	8.36
EC average[a]	10.03	8.63	9.43
CAP 1967/8 Target prices	10.625	9.125	10.30

a: weighted by production
Source: J. Marsh and C. Ritson, *Agricultural Policy and the Common Market*, European Series 16 (London, Chatham House, 1971)

CAP. (Table 20). At the outset, most CAP prices were set at a level above the average of the previous national prices despite warnings from economists that this would impose a cost on the EC in terms of lost potential output from resources retained in agriculture.[30] Farmers in France and Belgium benefited most from price increases for wheat and barley, whereas milk prices rose to the greatest extent in France and the Netherlands. Italy and West Germany were the two countries to gain the least from the new price levels. Those countries joining the Community in 1972 and 1981 have faced similar transitional adjustments to new price levels. For example, in 1971/2 the guaranteed price for barley in the United Kingdom was 26 per cent below the EC target price, while the equivalent price differences for wheat and beef were 24 and 19 per cent respectively. On the other hand, the national price of sugar beet was 6 per cent and the milk price only 2 per cent above the equivalent Community price. In the short term, producers throughout the EC received a real price-stimulus to increase production under the CAP, although the stimulus varied by product and country. In the longer term, increases in production costs under price inflation appear to have eroded these benefits (Table 9), but again in a variable manner by country and product.

A second source of variation in producer prices has emanated from the application of monetary compensatory amounts (MCAs). The impact of 'green' rates of exchange and MCAs in creating national variations in CAP

Table 21. Changes in target prices, 1972–6 and 1976–80[a] (in national currencies and in real terms)

Country	1972/3–1976/7 (indices)	1976/7–1980/1 (indices)
Ireland	120[c]	97
Denmark	98[c]	97
France	97	95
Belgium	98	92
United Kingdom	133[c]	91
West Germany	110	91
Netherlands	95	91
Luxembourg	100	90
Italy	116	89
Average[d]	108[c]	92

a: figures above 100 = price increase in this period; below 100 = decrease; b: converted through green rates and adjusted by the implicit GDP price index; c: allowing for accession adjustments; d: weighted by shares in total value of output 1974–6
Source: M. Tracy, *Agriculture in Western Europe* (Granada, London, 1982), p. 332

prices has been described already (Tables 8 and 10), while the cumulative effect has been to vary the prices received by producers of each commodity throughout the Community. When adjustments to new CAP price levels, changes in the relative value of national currencies, and the impact of national inflation rates are taken into account, the real value of product prices can be seen to have varied from commodity to commodity and from country to country (Tables 21 and 22). Wheat prices, for example, have risen to the greatest extent in the Netherlands, France and West Germany under the CAP, whereas Denmark, Italy and West Germany have gained the highest price increases for milk. Price ratios between alternative or competing products have also been altered within each Member State. In the Netherlands, for example, the prices of sugar beet, cattle and pigs have increased in relation to milk and wheat, while in Ireland price ratios have moved in favour of cattle and pigs at the expense of products such as wheat and sugar beet.

Table 22. National product price trends 1957–75 (1957/8 = 100)[a]

Country	Wheat	Cattle	Pigs	Sugar Beet	Whole Milk	All prices[b] (annual av. %)
United Kingdom	106	144	115	104	112	+2.5
Ireland	107	150	131	106	n.d.	+1.9
Italy	101	169	134	163	175	+0.4
West Germany	115	200	156	114	164	+0.1
Denmark	n.d.	199	145	124	229	−0.6
France	134	n.d.	n.d.	136	n.d.	−1.0
Belgium/ Luxembourg	104	184	154	136	156	−1.3
Netherlands	135	173	173	184	149	−1.9

a: weighted average prices received by farmers for all types & grades; b: expressed in real terms and in national currencies, 1972–80
Source: J. S. Marsh and P. J. Swanney, *Agriculture and the European Community* (London, George Allen & Unwin, 1980), Table 7; European Communities: *Green Europe Newsletter*, 13 (1980), Table 4.2

Thus new patterns of national advantage and disadvantage in the production of the various agricultural products have emerged, but in a complex way. Not surprisingly, no general, clear-cut relationship can be established for all Member States, whether as The Six or The Nine, between production trends and any single factor such as production costs or product prices under the CAP. Rather, the balance of costs and prices for the various products within each national agriculture has to be examined. Such an approach reveals that similar national production trends (Table 16) have emerged under varying national combinations of prices and costs. The role of the CAP, therefore, must be seen as enabling rather than causative for national agricultural trends in the EC.

Conclusion

Gradually, CAP prices have lost their real value during the last decade. Nevertheless, they have remained substantially above those ruling for the world market while the intervention system has provided an assured market for the major agricultural products of the Community. Together, these two aspects of the CAP have acted as a stimulus to the application of new farming technology, resulting in an expansion and intensification in the output of most agricultural products. An examination of the ratio of factor inputs to the total volume of agricultural production, for example, reveals increases in productivity of between 1.7 and 1.8 per cent a year since 1963 for the EC as a whole.[31] This is slightly above the 1.5 per cent assumed in the 'objective' method for setting agricultural price levels (see p. 71). National growth rates in the intensification of production have been highest in West Germany, Ireland and the Netherlands, and at their lowest in Denmark, the United Kingdom and Italy. Within each country, regional variations in increased production tend to reflect the location of products such as poultrymeat, sugar beet and pigmeat. (Figure 7b). The output of these products has increased greatly under the CAP and there is enormous potential for further increases in production with the application of modernised farming techniques throughout the Community.

The changing national concentration of agricultural production under the CAP has had ramifications for national levels of self-sufficiency of the various crops and livestock products. Faced by declining or stable levels of domestic consumption, the increased output of many products has become surplus to market demand (Table 23). However, the location of the surplus production varies from product to product. The Community's major problem product, milk, for example, is in greatest national surplus in Denmark, Ireland and the Netherlands, and decreasing levels of self-sufficiency are displayed only by Denmark and Belgium/Luxembourg. The EC sugar surplus is associated with high levels of self-sufficiency in Belgium/Luxembourg and France, but all countries have experienced rising levels of self-sufficiency since 1971. By contrast, lower and decreasing levels of self-sufficiency are evident in a majority of countries for fresh vegetables, fresh fruit, eggs and pigmeat. Overall, however, countries with the greatest national surplus of a product tend also to be those in which production is concentrating under the CAP (compare Tables 15 and 23). This tendency is to be expected within a common market where production is relocated in areas of greatest economic advantage. In the EC, however, the increased output of some countries is not sufficiently compensated by the declining production of others. The overall result is one of rising national levels of self-sufficiency and the further growth of production that is surplus to the requirements of the Community. No government seems prepared to accept the decline of individual sectors of its national agricultural industry. Indeed, in some cases, as with dairying in the United Kingdom, governments have actively encouraged the development of a product already in surplus in the Community although not in surplus in the national economy.

Table 23. National levels of self-sufficiency, 1980 (%)

Commodity	WG	F	I	N	B/L	D	Ire	UK	The Nine
Wheat	106	205	82*	61	74	126*	49*	80	116
Barley	91	178*	37	45*	80	111	117	116	113
Maize grain	23	140*	65	1	3	–	–	–	62
Potatoes	86*	107	100	135	95*	100	101*	95*	101
Sugar	125	200	93	156	247	187	113	46	125
Fresh vegetables	34*	95*	119*	191	113*	69*	89*	77	96
Fresh fruit[a]	53	97*	128	50*	60*	48*	22*	32*	79
Wine	45*	104	137	–	3*	–	–	–	105
Oils and fats	110	74	73*	98	92	125*	66*	37	81
Beef and veal	106	112	62	143	107	356	578*	78	100
Pigmeat	87*	84*	74*	243	159	368*	133*	63*	101
Poultrymeat	62	128	99	294*	88*	230*	98*	100	105
Milk (whole)	120	121	75	246	99*	222*	239	74	222
Eggs[b]	72*	101	95*	290	136*	102*	81*	99	102
Sheepmeat	38*	78	64	287	20*	–*	144	61	67

WG: West Germany; F: France; I: Italy; N: Netherlands; B/L: Belgium and Luxembourg; Ire: Ireland; UK: United Kingdom
a: excluding citrus; b: butter 119; cheese 105; SMP 115*; condensed milk 168; *: decreasing self-sufficiency, 1971–80
Source: Commission of the European Communities, *The Agricultural Situation in the Community, 1975 Report* (Brussels, 1976), pp. 187–190; 1982 Report, 1983, pp. 244–5

In addition, attempts to contain the problem of surplus production merely by manipulating CAP prices or production quotas at the EC-level seem doomed to failure. This chapter has shown that national production trends are influenced by factors other than product prices:[32] for example by the inflation rates in national production costs, variations in 'green' rates of exchange, and differences in the adoption of new farming technology. Moreover, the price ratios between alternative or competing products vary from country to country, while producers react differently over time to similar price changes.[33] Consequently, changes in market regulations or product prices introduced for the EC as a whole have had, and will continue to have variable and largely unpredictable consequences for each Member State.[34] Economists have advanced a number of options for dealing with the existing and prospective production of agricultural surpluses within the Community, but discussion of them is deferred until Chapter 11.

Notes

1 Developments in agricultural technology are described further in: M. Eddowes, *Crop production in Europe* (London, Oxford University Press, 1976).

2 J. Tuppen, *The economic geography of France* (London, Croom Helm, 1983), pp. 65–6.

3 The analysis in Chapter 6 excludes Greece, in part because that country was not in the Community for most of the time-period being considered, and in part because comparable data have not yet been published. A variety of measures of production trends have been used (value, volume, crop area/livestock number) depending on the availability of data and the comparability of the measures between the Member States. The difficulties of establishing time series for even two countries – West Germany and the United Kingdom – are examined by: D. Andrews *et al*, *The development of agriculture in Germany and U.K. Vol. 3: Comparative time series 1870–1975*, Miscellaneous Study 4 (Centre for European Agricultural Studies, Wye College, University of London, 1979).

4 Only Italy has a higher proportion of agricultural output from crops (62 per cent) compared with livestock (38 per cent).

5 G. Weinschenck and J. Kemper, 'Agricultural policies and their regional impact in Western Europe', *European Review of Agricultural Economics*, VIII (1981), pp. 251–81.

6 D. M. Phillips, 'West German study sees European Community grains sufficiency soon', *Foreign Agriculture*, XII (1974), pp. 6–9.

7 M. L. Debatisse, *E.E.C. organisation of the cereals markets: principles and consequences*, Occasional Paper 10 (Centre of European Agricultural Studies, Wye College, University of London, 1981), p. 26.

8 House of Lords Select Committee on the European Communities, *Imports of cereal substitutes for use as animal feedingstuffs*, 42nd Report, HL270 (1980/1).

9 I. Smith, 'Europe's sugar dilemma', *Journal of Agricultural Economics* XXXI (1980), pp. 215–23.

10 Centre for European Agricultural Studies, *The E.E.C. milk market and milk*

policy, Seminar Paper 3 (Wye College, University of London, 1977), p. 24.

11 A similar study for The Six, using data for the period 1958–73, revealed broadly similar trends. Attention is drawn to those countries and commodities where variations in trends between the time periods 1958–73 and 1969–81 are evident.

12 R. H. Best, 'Land use structure and change in the EEC', *Town Planning Review* L (1979), pp. 395–411.

13 The concentration of different fruit and vegetable crops is discussed by: L. Hinton, *The European fruit and vegetable sector*, Occasional Paper 20, Department of Land Economy, (1976) Cambridge University.

14 C. E. Goldthwaite, 'Expansion era marks German wine industry despite 1980s downturn', *Foreign Agriculture* VIII (1980), pp. 18–20.

15 A. J. Rayner *et al*, 'A time-trend analysis of sugar and cereal yields in the EEC', *Oxford Agrarian Studies* XI (1982), pp. 1–47.

16 Rayner, *Oxford Agrarian Studies* XI (1982), pp. 1–47.

17 Smith, *Journal of Agricultural Economics* XXXI (1980), pp. 215–23.

18 M. Tracy, *Agriculture in Western Europe: challenge and response 1880–1980* (2nd ed. London, Granada, 1982), p. 336. There are widely different regional cost structures for the production of sugar beet. Without quotas, low-cost producers in Belgium and France would probably capture the market at the expense of producers in other countries.

19 Centre for European Agricultural Studies, *Survey of the potato industry in the EEC,* Report 8 (Wye College, University of London, 1975).

20 The point biserial correlation coefficient was calculated for the relationship between the degree of national concentration of a product in 1969 and the subsequent trend in concentration (increase or decrease). A positive coefficient indicated greater concentration in countries already accounting for most of the product, whereas a negative coefficient showed the product to have become more dispersed amongst Members of the Community.

21 R. Behrens and H. Haen, 'Aggregate factor input and productivity in agriculture: a comparison for the EC member countries 1963–76', *European Review of Agricultural Economics* VII (1980), pp. 109–46.

22 National Economic Social Council, *A comparative study of output, value-added and growth in Irish and Dutch agriculture,* Publication 24 (Dublin, Stationery Office, 1976).

23 The highly productive areas of agriculture in West Germany are the Börde Loess region/Lower Rhine Valley/Wurzburg basin (favourable climate and soils), Niedersachsen (large farm structure), and the Elbe Valley around Hamburg/areas around Munich and Nuremberg-Fürth (intensive agriculture) and are detailed by: D. Burtenshaw, *Economic geography of West Germany* (London, Macmillan, 1974), p. 166.

24 The cost advantage was measured by the amount of resources of land, labour and capital needed to produce a given volume of agricultural output, and is presented

in: D. K. Britton, *The development of agriculture in Germany and U.K. Vol. 4: A comparison of output structure and productivity*, Miscellaneous Study 5 (Centre for European Agricultural Studies, Wye College, University of London, 1981), p. iii.

25 House of Lords Select Committee on the European Communities, *State aids to agriculture*, 7th Report, HL90 (1981/2), p. xxvii and pp. 154–5.

26 Andrews, *The Development of agriculture in Germany and U.K.* (1979), p. 38.

27 Behrens and Haen, *European Review of Agricultural Economics* VII (1980), pp. 109–46.

28 C. J. Doyle, 'A comparative study of agricultural productivity in the United Kingdom and Europe', *Journal of Agricultural Economics* XXX (1979), pp. 261–70.

29 Agriculture Economic Development Committee, *Milk production in the European Community: a comparative assessment* (London, National Economic Development Office, 1981).

30 J. Marsh and C. Ritson, *Agricultural policy and the Common Market* (London, PEP, 1971), p. 34.

31 Behrens and Haen, *European Review of Agricultural Economics* VII, 1980, pp. 109–46.

32 Centre for European Agricultural Studies, *The EEC milk market*, p. 30.

33 Centre for European Agricultural Studies, *The EEC milk market*, p. 27.

34 Debatisse, *E.E.C. organisation of the cereals markets*, p. 31.

7
Regional trends in the specialisation of agriculture: crops

Agricultural production has always been highly diversified between and within Member States of the European Community. Each country and region has experienced a different economic and social history of development including variations in industrialisation, urbanisation, land inheritance, trade policy, consumer preferences and domestic price policies.[1] In addition, the great areal extent of the EC, covering wide variations in latitude, longitude and altitude, has ensured a diversity of physical conditions for agricultural production. Farmers have responded to variations in soils and climate as well as to factors such as nearness to urban markets in developing their systems of production. Although these features together have produced a mosaic of farming types in Western Europe,[2] many countries and regions have become associated with a few particular agricultural products.

Regional descriptions of agriculture in individual countries of the EC are widely available.[3] Here, by contrast, regions of the whole Community are considered together. Attention is focussed on changing patterns of agricultural specialisation under the CAP using a product-by-product rather than region-by-region approach. There are, however, considerable difficulties in assembling consistent regional data for even one agricultural product,[4] and the problems are magnified if changes over time are investigated. Previous studies of regional specialisation, for example, have been confined to one time period, to a few major land uses and livestock products, and to The Six rather than The Nine.[5] In this chapter regional data for eight major agricultural products have been abstracted from a recent study by an international group of agricultural economists.[6] The data have been subjected to further statistical analysis and the conclusions drawn by the author are not those of the study group. The data cover the period 1964/5 to 1976/7 and they have been supplemented by reference to maps of agricultural change for the years 1951/5 and 1969/71 published by Eurostat.[7] All the data sources are based on eighty administrative regions within the Member States.

Specialisation has been measured as the proportion of the total value of agricultural output of a country or region accounted for by a particular product. By comparing data across all regions for two time periods, an index of the changing regional specialisation of the product can be derived. Clearly, the trends towards concentration (Chapter 6) and specialisation in agriculture are inter-linked:[8] a region which is becoming more specialised in a particular product is also likely to account for a greater share of the total output of that product. However, by focussing on specialisation at the regional level, the problem posed by the varying sizes (resource bases) of the regions is removed. Figure 8 provides a Community perspective to the analysis by showing the changing degree of specialisation of each national agriculture between 1963 and 1981.

Cereals
Priority must be given to cereals in any consideration of agriculture in the EC. Grains are an important cash crop on a large number of farms, but in addition they form a significant input to many livestock enterprises. About 35 per cent of cereals produced in the Community are employed in compound animal feeds and a further 25 per cent are fed on the farm. It was not by accident, therefore, that cereals were the first commodity to achieve common prices under the CAP. So central are cereals to farming in Western Europe that they are found on over 60 per cent of holdings in the Community; national proportions range from a maximum of over 90 per cent in Denmark to a minimum of 22 per cent in the Netherlands (Table 24). Specialist cereal farms, however, are important only in Denmark, although they account for nearly 10 per cent of all farms in Italy.

An increase in the area under cereals has been a major feature of agricultural change in the Community under the CAP. However, the degree of specialisation in cereals by the Member States has tended to diverge over the last two decades. Cereals account for an increasing proportion of national output in West Germany, France, Denmark and the United Kingdom, whereas Italy, Belgium and the Netherlands have become less dependent on grain crops. France has experienced the greatest rate of increase in specialisation, and Italy the greatest decrease. Today the major regions of cereal specialisation lie in northern and south-western regions of France

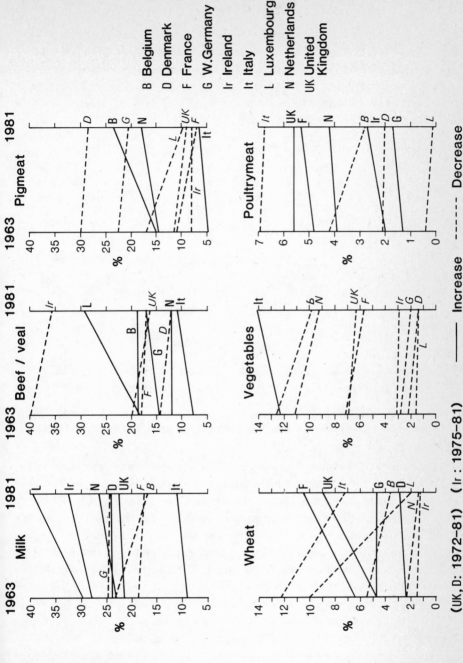

(UK, D: 1972–81)   (Ir: 1975–81)   —— Increase   ----- Decrease

Figure 8. Changing national specialisation in agricultural production, 1963–81 (volume)

B Belgium
D Denmark
F France
G W.Germany
Ir Ireland
It Italy
L Luxembourg
N Netherlands
UK United Kingdom

together with eastern areas of England, Denmark, and isolated regions in southern Italy (Molise, Basilicata – Figure 9a). In Centre and Ile-de-France, on the rich limon soils of the Paris Basin, for example, cereals account for half the value of agricultural output; similar levels of regional specialisation are reached only by the Midland and Western regions of Ireland in beef production and by the Languedoc in wine.

Table 24. The structure of cereals production

Country	% holdings with cereals	Ave. ha cereals/holding	Cereal farms as % total farms[a]	UAA/AWU[b] on cereal farms (ha)
West Germany	84	7.2	4.2	16.8
France	66	12.0	5.2	35.7
Italy	55	3.3	9.3	11.8
Netherlands	22	6.8	0.3	22.6
Belgium	57	5.6	1.3	11.4
Luxembourg	78	8.9	1.4	11.9
United Kingdom	42	32.0	6.7	36.2
Ireland	40	4.0	1.2	24.4
Denmark	92	15.0	20.7	23.0
EC	61	7.6	7.0	22.1
(Greece)	(52)	(1.7)	(n.d.)	(n.d.)

a: 1975 Farm Structure Survey; b: UAA – utilised agricultural area, AWU – annual work unit
Source: Commission of the European Communities, *The Agricultural Situation in the Community, 1981 Report* (Brussels, 1982), p. 291 and 299. *1982 Report* (1983), p. 313

Increases in the regional specialisation of cereals have been greatest in the main areas of production, namely central and northern France, eastern and northern parts of Britain, Denmark, and parts of southern Italy (Table 25 and Figure 9b). Elsewhere, high proportional increases in specialisation are generally associated with areas where cereals have increased from a small initial area of production (Highlands of Scotland, Baden-Württemberg). In general, there is evidence of a northward shift in the production of cereals, a trend emphasised by the declining share of cereals in the regional production of most parts of central and northern Italy. Only Molise, Basilicata, Sicily and Sardinia show significant increases in cereal production in the Mediterranean regions of the EC in the last two decades.

Yields per hectare of all cereals have risen in recent years, in part through the development of higher yielding varieties, but also through the increased application of artificial fertilisers. Between 1976 and 1981 alone, cereal yields in the EC increased by twenty-five per cent. The highest national yields for cereals occur in Denmark and the Netherlands and the lowest in Italy. Nevertheless, the greatest rates of increase in yields are evident in southern Italy, central and southern France, the west of England and Ireland. Should average yields be raised throughout countries such as France just to the average for the EC, even greater surpluses of grain would be generated.

The regional balance of production of the different cereals has been changing under the CAP. Wheat, for example, accounts for forty per cent of cereals production in the Community and contributes most to the

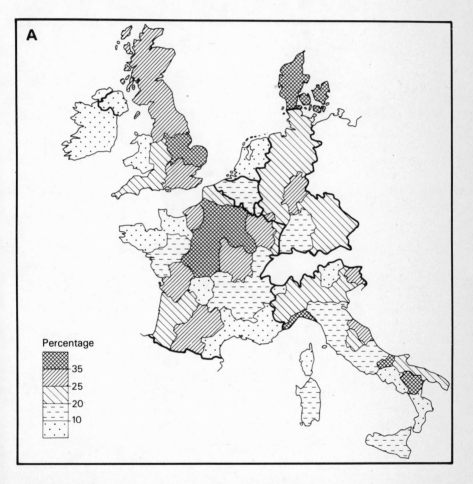

Figure 9A. Cereals as a percentage of the total standard gross margin of the region, 1975

agricultural output of France (Centre, Picardy, Champagne, Burgundy) and the United Kingdom (eastern England, Scotland) (Figure 8). Within the same countries, however, production has been declining in southern and north-western France, Wales and northern England. Generally, the rising area under wheat has been at the expense of other cereals such as rye in the case of West Germany. In Denmark, on the other hand, the increasing regional specialisation of wheat in eastern Jutland and the islands of Zealand and Funen has been at the expense of pasture for dairy cows.[9] The expansion of wheat was encouraged by the separation of bread and feed wheat prices

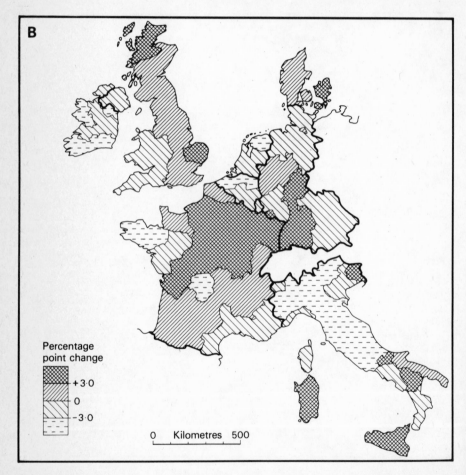

Figure 9B. Changing percentage of cereals in the total agricultural production of the region, 1964/5–1976/7

under the CAP in 1977:[10] bread wheats were awarded a higher reference price. Almost all wheat is now marketed as bread wheat even though the quality is not good enough for milling unless improved by imports of wheat from North America. Thus, despite increasing domestic production, imports of high quality wheat are continuing. Not all countries, however, are participating in the increasing production of wheat: the crop accounts for a falling proportion of total output in countries such as Italy (except in the south), Luxembourg, and Belgium (Figure 8).

Barley comprises 34 per cent of cereals production in the EC but is more regionally concentrated than wheat. The crop is important in only three countries – the United Kingdom, Ireland and Denmark – although it is grown widely in each. Indeed, in the United Kingdom the crop has become more dispersed rather than localised in its distribution.[11] In Ireland and Denmark, barley is still more important than wheat in the national economy, but in the United Kingdom these positions have recently been reversed. In France, barley is associated with northern regions such as Centre, Burgundy and Champagne, and with Brittany and Poitou-Charentes in the west. These areas have experienced the greatest increases in production so heightening the regional contrast between northern and southern parts of the Community as far as cereal production is concerned. The only regions with decreases in barley production are in southern Italy, Sicily and Sardinia.

Maize grain has achieved a prominent agricultural position in France and Italy. In the early 1950s the crop was associated mainly with northern Italy, but in later years it was adopted by farmers in south-west France (Aquitaine, Gascony). The development of drought-resistant and early-maturing varieties in the 1960s led to the diffusion of maize north of the Loire while traditional arable farms – especially south and east of Paris – were able to apply their expertise and capital-intensive methods to the crop.[12] Even so, the role of co-operatives, such as *Limagrain* in the *département* of Puy-de-Dome,[13] was influential in providing financial support and marketing expertise in the diffusion of the crop. At the same time, the area under maize grain has declined throughout Italy, and only increasing yields have produced a rising volume of production, especially in northern areas such as Veneto and Piemonte.

Maize has also increased in popularity as a silage crop for feeding to livestock such as dairy cows and beef cattle. The crop is harvested while green to form a high energy, digestible, bulk feed. In West Germany, for example, 85 per cent of the area under maize is harvested for silage while over 40 per cent of the maize grain crop is fed to livestock on the farm on which it is produced. In France, the area under maize silage rose from 279,000 hectares in 1970 to 1.1 million hectares in 1979 and is now an established feature of agriculture in regions such as Brittany and Pays de la Loire. In Italy, maize silage has been used intensively in beef fattening since the mid-1960s on the lower plain of Lombardy (Cremona, Mantova),[14] often at the expense of maize grain.

Other cereal crops, such as oats and rye, retain an importance in only a few regions of the EC. Their production has been declining and yields have not increased to the extent enjoyed by wheat and barley. In the United Kingdom, for example, oats retain a hold mainly in the cool, moist uplands of Wales and Scotland, while rye is of importance in northern parts of West Germany and in scattered localities within Belgium and Denmark.

Fresh fruit and vegetables
The importance of fruit and vegetable production has always been highly variable within the countries of the EC. Fresh vegetables, for example, are of national importance in just three countries – Italy, the Netherlands and Belgium – but only in Italy have vegetables comprised a rising proportion of total agricultural output (Figure 8). Elsewhere in the EC, vegetables have declined relatively in importance in recent years. For fresh fruit, convergence is taking place on a range of values from 0.4 (Ireland) to 3.3 per cent (France) of total national output. Again, only Italy with the advantage of a Mediterranean climate retains a significant, if falling, degree of specialisation in fruit (7.7 per cent total output).

In contrast to cereals, fresh fruit and vegetables are produced under minimal price support regulations: a low common tariff operates for all imports, but the withdrawal of produce from the market has been available only for cauliflowers, tomatoes, peaches, table grapes, pears, apples, mandarins, oranges and lemons for the period under study. Producer organisations (often regional co-operatives) operate the withdrawal system to prevent prices in the market falling too far below the reference prices for the various crops.[15] Payments from the Agriculture Fund in support of the system have gone mainly on tomatoes, citrus fruit and apples. More recently 'penetration premiums' have been paid to assist the marketing of Italian citrus fruit, import certificates have been imposed for 'sensitive' processed products, and processing aids provided for tomato products, canned peaches and prunes. Nevertheless, in comparison with other agricultural products, regional variations in production costs have had a more pronounced impact on the changing location of fruit and vegetable production.

Table 25. The changing regional specialisation of production within Member States (1964/5–1976/7)

Country	Cereals	Sugar beet	Milk	Oils & fats	Beef & veal	Intensive livestock[b]	Wine	Fruit & veg.[c]	Other[d]
France	+2.8	−0.1	−2.4	−0.1	+1.2	−3.1	+2.6	+0.5	+0.3
Italy	−1.7	+0.4	+0.9	+0.4	+1.8	+1.1	−3.1	−0.5	+0.3
Belgium	−3.0	+1.3	−4.5	+0.9	+4.2	+2.7	−	+0.7	−1.9
West Germany	+0.7	−0.1	−0.5	+0.2	+2.5	−3.6	+3.9	+0.4	−1.9
Netherlands	−1.4	−1.5	+0.4	+0.3	−2.5	−0.6	−	−1.6	+6.9
United Kingdom	+1.5	−0.8	+1.8	−	+0.2	−3.9	−	−1.8	+2.2
Ireland	−2.6	−0.5	+2.9	−	+7.3	−5.2	−	−	−1.9
Denmark	+4.8	+1.1	+2.9	−	−0.6	−8.5	−	+0.1	−0.3
EC	+0.3	−0.3	+0.02	+0.2	+1.8	−2.2	+0.04	−0.3	+0.3

Positive values denote increasing and negative values decreasing overall specialisation within the regions of each country[a]
a: Σ (% total regional output by a product 1976/7 − % total regional output by a product 1964/5) ÷ Number of regions in the Member State; b: Pigs, poultry, eggs; c: subject to CAP regulations; d: all products not subject to CAP regulations by 1976
Source: author's calculations from RICAP data[6]

At a regional level, it is necessary to emphasise the variety of farming systems used for the production of fresh fruit and vegetables. Certainly, all systems have in common high capital investment and high intensity of output per hectare and per worker. But they vary in character from the heated glasshouses in Boskoop in the Netherlands, through the small, intensively-farmed plots of southern Italy, to the large mechanised arable fields of East Anglia and the extensive orchards of southern France. With these variations in mind, two areas can be identified within the EC where fruit and vegetables are significant elements of the regional agricultural

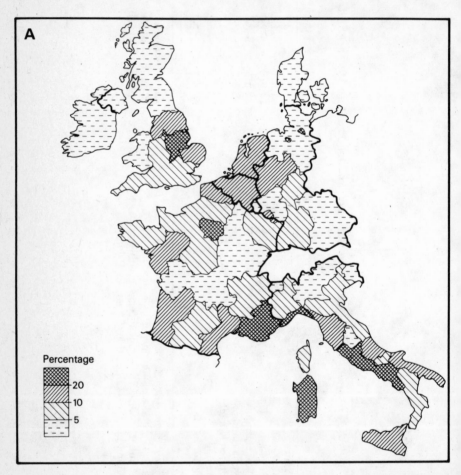

Figure 10A. Fruit and vegetables as a percentage of the total standard gross margin of the region, 1975

economy (Figure 10A). One area stretches in an arc around the Mediterranean coast of the EC from Roussillon in the west, throught Côte d'Azur and Liguria, to Lazio and Campania in the east. A relatively high proportion (a quarter) of the produce of this area is subject to CAP regulations, especially in southern Italy, but fruit and vegetables not subject to price supports also characterise agricultural production. The greatest regional specialisation occurs in Provence-Côte d'Azur and Campania where the fruit crops include citrus, peaches and table grapes. In the former area, production occurs mainly in the lower Rhône Valley under irrigation and

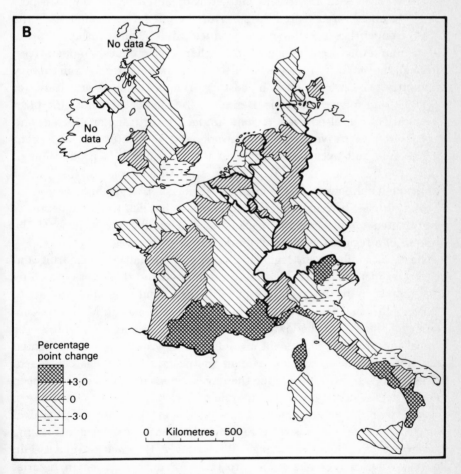

Figure 10B. Changing percentage of fruit and vegetables in the total agricultural production of the region, 1964/5–1976/7

increasingly under plastic greenhouses. This last development offers cheap and flexible protection to vegetable crops, especially in the small market gardens, with improved yields and earlier harvesting dates.

The second area of specialisation occurs in northern parts of the EC in regions adjacent to the major urban conurbations: eastern England, Belgium, the Netherlands and Nordrhein-Westfalen in West Germany. Compared with the Mediterrranean area, a lower proportion of fruit and vegetable production is subject to CAP market regulations. Typical are the economies of scale enjoyed by glasshouse producers in the Netherlands which, together with an efficient marketing organisation and low transport costs to market have ensured the survival of crops such as tomatoes in the face of competition from lower-cost producers in Italy and southern France. Often, marketing organisation is in the hands of producer-co-operatives, although the proportion of sales handled by them varies greatly from country to country. The proportion is highest for fruit in Denmark and Italy, for example, and for vegetables in Denmark and the Netherlands. Producer co-operatives can also have a regional impact. In Brittany an efficient and aggressive co-operative marketing network has concentrated over 70 per cent of France's and over one-quarter of the Community's production of cauliflowers in the region. Regionally, specialised crops in the United Kingdom (cauliflowers in the Holland region of Lincolnshire, carrots in Norfolk, apples in Kent), by comparison, have not had the same degree of co-ordinated quality control, packaging and production because of the absence of a regional network of marketing co-operatives.

The data available on regional changes in specialisation in fruit and vegetables distinguish between products covered and those not covered by CAP marketing regulations. In Figure 10b, fruit and vegetables subject to CAP regulations show a shift of emphasis towards certain Mediterranean regions of the EC. In the lower-Rhône/Languedoc region, for example, the radical but localised movement away from vine-growing towards irrigated fruit and vegetables, already evident in the mid-1960s[16] has continued. Large-scale regional schemes for the development of irrigation have been central to these changes,[17] but the role of agricultural extension workers, together with the establishment of a network of experimental farms to deal with the cultural problems of the region, have been important in facilitating the process of agricultural change. Within Italy, the greatest increases in specialisation in products subject to the CAP have occurred in Liguria, Trentino (Adige Valley), Campania and Calabria, but attention is drawn to the decreasing specialisation which characterises eastern parts of Italy together with Sicily and Sardinia. At a sub-regional scale, 'CAP crops' have become more localised within these latter regions. In the province of Emilia-Romagna, for example, tomatoes have become localised in the 'agricultural zones' of Pianura Piacenza and Basso Alto.[18] All of these trends were evident before the operation of the CAP but have been intensified in recent years. In particular, the Mediterranean regions of the Community, which have a specialisation in intensive crops, have generally become more dependent on fruit and vegetables subject to CAP regulations compared with crops lacking such support.

Elsewhere in the Community production trends have been rather different. With the exception of Belgium, produce not subject to the CAP has become more rather than less important within total fruit and vegetable production, and only low levels of increasing specialisation in 'CAP crops' have been recorded (Figure 10b). This last development is evident throughout western France and West Germany with the development of top-fruit orchards such as apples, and in certain parts of those regions already specialised in fruit and vegetable production (East Anglia, northern Belgium, the Netherlands). Produce not subject to the CAP include potatoes, field crops such as celery, spinach and carrots, and flowers, plants and bulbs. Flowers and plants, for example, with their high value, account for 20 per cent of the market for horticultural products. In the Netherlands and Belgium, small horticultural holdings are responsible for a significant proportion of such production, 13 and 6 per cent of holdings respectively being classified as specialising in horticulture. Similar proportions of holdings farm under glass, although farms specialising in field crops (10 and 8 per cent respectively) are equally important for produce such as potatoes, cabbages and broccoli.

Statistical data and published studies for vegetable crops in the EC are sparse. Consequently, attention is turned to fruit crops and then to potatoes for a more detailed consideration of the regional specialisation of production. Farms specialising in fruit are significant only in Italy (especially Sicily, Emilia-Romagna, Campania, Piemonte), although the average area of fruit crops per farm tend to be small compared with production in the United Kingdom, Denmark and the Netherlands. France (Rhône-Alpes, Provence/Côte d'Azur) also emerges with a favourable structure of production when the average area of fruit crops farmed by each worker is examined (Table 26); in addition the proportion of holdings with these crops is relatively high compared with other countries.

Apples, followed by pears and peaches, account for the largest volume of

Table 26. The structure of fruit and berry production

Country	% holdings with fruit & berries	Ave. ha fruit & berries per holding	Fruit farms as % total farms[a]	UAA/AWU[b] on fruit farms (ha)	Golden Delicious apples as % total orchards	% total apple orchards in three varieties
West Germany	6.7	0.9	1.6	3.1	23	56
France	9.5	1.7	3.3	4.7	62	67
Italy	14.0	1.4	20.1	4.2	38	72
Netherlands	5.9	3.1	4.1	3.0	32	74
Belgium	4.2	2.3	4.2	2.9	48	82
Luxembourg	3.5	0.5	0.5	3.7	26	68
United Kingdom	3.9	5.5	1.7	5.0	3[c]	69
Ireland	0.3	1.6	0.1	5.8	39	67
Denmark	1.7	4.2	1.2	3.4	10	39
EC	10.0	1.6	10.5	4.2	37	59

a: 1975 Farm Structure Survey; b: as Table 24; c: Cox's Orange Pippin, 54%
Source: as Table 24 and European Communities, *Community Survey of orchard fruit trees*, Eurostat (Brussels, 1976), pp. 18–55

fruit production in the EC. Apple orchards illustrate many of the problems of fruit production including their age-structure and increasing dependence on a few varieties (Tables 26 and 27). But again, the scale of such problems varies from country to country and from region to region. Luxembourg and the United Kingdom, for example, have the highest proportion of apple orchards over twenty-five years of age, whereas amongst the three major producers, the youngest orchards are found in West Germany. Grant aid for grubbing-up orchards has been available under the CAP since 1971, and in the first two years of the scheme 54,000 hectares of apples and 21,000 hectares of pears disappeared. However, in subsequent years the increased output of younger trees tended to compensate for reductions in area.[19]

Table 27. Age structure of apple orchards in the EC (%)[a]

Country	Age (years)					% Golden Delicious less than 1 yr old
	below 5	5–9	10–14	15–24	over 25	
West Germany	13	31	28	19	9	24
France	10	26	39	21	5	36
Italy	16	20	22	23	19	53
Netherlands	6	23	34	20	16	23
Belgium	23	37	18	12	11	49
Luxembourg	3	16	15	6	60	n.d.
United Kingdom	12	16	16	26	30	n.d.
Ireland	40	22	8	16	15	n.d.

a: no data for Denmark
Source: as Table 26

The type of apple variety also influences the profitability and hence the competitiveness of orchards. Golden Delicious apples are the single most important variety in the Community (37 per cent of the total orchard area), giving regular cropping, large-sized fruit, high yields and good storage properties. Orchards in France, and to a lesser extent in Belgium are dominated by this one variety, and as few as three varieties account for over 65 per cent of the orchard area in most countries. Only Denmark and West Germany retain a range of apple varieties. In the United Kingdom, where the lower-yielding Cox's Orange Pippin is the main apple variety (Table 26), producers have been at a disadvantage in competition with imported Golden Delicious apples from France. Such imports increased by 25 per cent between 1970 and 1980. French apple orchards became more localised in Provence,

Côte d'Azur, Languedoc (and Pays de la Loire) during this time and benefited from production under a Mediterranean climate.

Table 28. Regional variations in apple orchards – Italy

Region	% irrigated	% Golden Delicious	% older than 25 years	% fewer than 400 trees/ha
Val Padana	51	30(31)a	17	64
Piemonte	58	74(16)	10	62
Alto Adige	79	46(10)	19	72
Centrale	47	10(39)	24	72
Meridionale	42	2(23)	35	67
Italy	61	38(22)	19	67

a: Red Delicious
Source: as Table 26

Italy exemplifies how these features can vary by region (Table 28). The oldest, and least productive, orchards are prominent in the Centrale and Meridionale regions and these areas are also least dominated by Red and Golden Delicious varieties. In addition, a lower proportion of the orchards in these two regions is irrigated compared with the other parts of Italy, and there are higher densities of trees to the hectare. The aggregate result is that producers in northern areas (the provinces of Cuneo, Bologna, Modena, Ferrara, Ravenna) have increased their specialisation and share of apple production at the expense of these regions with their lower yields and higher production costs.

National specialisation in the production of potatoes is greatest in the Netherlands and the United Kingdom, although the crop retains a significance in a small number of regions within each Member State. In Denmark, for example, potatoes are important in west and north Jutland, generally on small farms,[20] while in the Netherlands the crop is most significant on the clay soil areas in southern parts of the country. Regions vary in their emphasis on seed, early and maincrop potatoes. Seed potatoes, for example, are locally important in Brittany and Normandy in France, and in Lower Saxony in West Germany; early potatoes characterise production in Brittany, southern Italy (Campania, Puglia, Sicily), and the south of France (Provence, Languedoc, Pyrénées); whereas maincrop potatoes are predominant in eastern England, the Netherlands and Ireland.

The structure of production also varies. Over 60 per cent of holdings in West Germany produce potatoes, but the average area of the crop is only 0.7 hectares on each farm. In the Netherlands and the United Kingdom, by contrast, there are fewer but larger producers – 29 and 23 per cent of holdings respectively have potatoes, with an average of 3.5 hectares per farm. In eastern England, as well as north-east France, potatoes are part of a large-scale, mechanised crop rotation that can include peas, maize, sugar beet and cereals. By contrast, in the south of France production is on a market-garden scale.

Potatoes are produced without benefit of CAP market regulations, and the sharply declining area of the crop, especially maincrop potatoes, has become more localised in regions already specialising in production. Increasing proportions of output are now taken for the production of starch – 40 per cent of production in the Netherlands (Drenthe, Groningen) – and for processing into chilled and frozen chips, crisps and powdered potato. In the Netherlands, 12 per cent of the crop is processed, while West Germany has experienced a 17 per cent annual rate of increase in the throughput of processed potatoes in recent years (15 per cent of the crop is processed). The trade in potatoes reflects these variations in the nature and location of potato production. Seed potatoes, for example, are exported from northern West Germany to Portugal and North Africa; early potatoes from the Mediterranean regions now compete in northern markets, especially in West Germany, with locally-grown produce; but the Netherlands remains the major exporter of seed and maincrop potatoes both to other Members of the Community and to countries outside the EC.

Wine
The regional localisation of wine in the EC will probably be more appreciated by the general reader than any other product. Distinctive associations exist between particular types and qualities of wine and small areas of production: the Gamay grape interacts with local granitic soils to produce the distinctive wine of Beaujolais, for example.[21] But wine is an infinitely varied product and a distinction must be drawn between the poor-quality table wine (73 per cent of EC production) that might be produced by a small farmer, and the higher-quality wines of a regional co-operative or large château. The latter have been subdivided since 1970 under the CAP[22] into quality wines psr (produced in specified regions), sparkling and liqueur wines. In West Germany, for example, 46 per cent of wine is handled by co-operatives, 25 per cent is sold to merchants, and a further 14 per cent is processed and bottled by the producers; the remainder is sold as grapes.[23] The majority of wine is classified as quality rather than table wine (Table 29). In Italy, by contrast, the production of table wine predominates.

The Community produces approximately 48 per cent of the total world production of wine from 2.7 million hectares of vineyards (27 per cent of total world area). Vineyards are farmed throughout the Member States, even in northern countries such as the United Kingdom. However, excluding Greece,[24] there are four broad areas where wine is a major contributor to the total value of agricultural output.[25] Two areas lie within France, and they are among the world's most important vineyards, whether measured in areal extent, volume, value or quality of output. In one area in the south of France – Rhône/Languedoc, Côte d'Azur – wine accounts for between 40 and nearly 70 per cent of a region's agricultural output. The Midi alone produces nearly 13 per cent of the world's wine. On the Languedoc plain vines are the predominant land use but they yield mainly low-quality table wines (vin ordinaire). The second area is comprised of more dispersed vineyards throughout west and south-west France (Aquitaine to the Loire Valley) where higher-quality wine is produced. A third region is centred on central Italy (Tuscany, Lazio/Abruzzi), although vineyards occur throughout the country and are locally important in areas as diverse as Sicily and Veneto. In Italy, 46 per cent of farm holdings have vines (Table 29), but the marked regional dependence on wine, as evident in parts of France, is largely absent. The fourth focus of wine production in the EC is located on the hills overlooking the Rhine and Mosel valleys of West Germany, usually on the south-facing valley slopes. But to this area should be added the adjacent vineyards of Luxembourg and Alsace, as well as those in Champagne and Burgundy in France. The vineyards are quite localised and in West Germany fewer than 10 per cent of all holdings are involved in producing wine.

When changes in the regional specialisation in wine production are examined, a distinction must be drawn between the volume and the value of

Table 29. The structure of wine production

Country	% holdings with vines	Ave. ha vines/ holding	Vineyards as % total farms[a]	UAA/AWU[b] in vine- yards (ha)	Table wine as % total wine
West Germany	7	1.5	3.8	2.4	4.3
France	38	2.5	12.9	5.8	64
Italy	46	1.1	10.5	3.4	87
Luxembourg	19	1.2	13.2	1.4	33
EC	31	1.4	8.3	4.4	73
(Greece)	(32)	(0.5)	(–)	(–)	(92)

a: 1975 Farm Structure Survey; b: as Table 24
Source: as Table 24

production. Taking volume first, the Italian regions have increased production to a greater extent than those in France and West Germany since the late 1950s, and now account for 53 per cent of production in the EC (France 44 per cent). Nevertheless, the area of vines has been in decline both in Italy and France. When the value of production is considered, however, the Italian regions show a declining specialisation in wine, especially in Puglia, Basilicata and Trentino. By contrast, French (Côte d'Azur) and West German (Rheinland-Pfalz) regions exhibit the contrary trend (Table 25) of increasing specialisation. At issue is the low-value, low-quality wine which characterises production in Italy, and the higher-value, quality wines of France (outside the Midi) and West Germany. These regional differences have been a continuing source of friction between Italy and France. Imports of cheap Italian wine into France have resulted in a long-standing 'wine war'; this erupted in 1975, and again in 1980, with the physical and sometimes violent blockade of imports by French producers,[26] and the imposition of an illegal tax on wine imports by the French government.

Perhaps more than any other single product, the changing geography of wine production is closely regulated by the CAP. Wine is a source of employment for nearly three million people in the EC, including its production, conversion and marketing. In addition, the product is a central and unifying element underpinning the socio-economic fabric of many regions where it is produced. The regional importance of wine is reflected in a range of policy measures that are highly complex, subject to constant revision, but which have evolved mainly since 1970 when they replaced equally complex national regulations. As with several other products, the central problem is the increasing output of wine (0.56 per cent a year) that is surplus to demand in the Community – an average of 5 per cent of production each year. But additional difficulties are posed by the falling per capita consumption of wine in the EC and by marked yet unpredictable fluctuations in yields from year to year. For instance, 1979 was a year of particularly high output in The Nine (176 million hectolitres), whereas production was unusually low in 1974 (128 million hectolitres). The intervention system helps to guarantee a minimum income for producers of table wine, but since 1970 surplus production has been distilled, the proportion rising in recent years to between 9 and 14 per cent of annual output. Not only have payments from the Agriculture Fund increased sharply, but instability has been created 'downstream' in the market for industrial alcohol.

Consequently, since 1976 a 'structural' rather than a 'price' policy has been developed for wine. Limitations on the planting of new areas of vines for table wines, grant aid to convert (i.e. grub-up) vineyards to alternative crops either temporarily or permanently, and schemes to replant vinyards with quality-wine stock have all been implemented. These structural policies have been set in the context of grant aid to restructure the vineyards of the EC which are notoriously small in size (an average of 1.4 hectares) and fragmented in distribution (Table 29). Only 30 per cent of holdings producing wine, for example, contain more than one hectare of vines, and of these 96 per cent have fewer than five hectares of vines. In addition, the great diversity of the vine-growing areas has been recognised by the regionalisation of the regulations covering production. Following Regulation 337/79, the 'Action Programme' of 1980–6 recognises five distinct regions[27] based on a 'natural suitability' for wine production (environmental and climatic). Planting in some zones is limited to quality wines, whereas in others financial aid to cease production is available.

The contraction of wine production, already evident in France under national policies,[28] has been fostered under the CAP. In the Languedoc, for example, the conversion of vineyards has been a feature of areas peripheral to the *plaine viticole*, especially in the south west. This process has continued a retreat of vines from the hill areas which was begun by the replanting of vineyards in the late-nineteenth century after the phylloxera epidemic. But even within the principal wine-producing *départements* – Gard, Hérault and Aude – higher rates of conversion can be identified in the vicinity of certain towns such as Montpellier, St. Pons and Lunas.[29] Such variations are a product of a number of processes which include the sale of land for urban development, the availability of irrigation for alternative crops, and the withdrawal of farmers from wine production for non-farm employment or old-age retirement. In less favoured areas, such as those between Vienne and Valence in the Rhône Valley, with their steep slopes and fragmented vineyards, production is on the decrease with the survival of only the higher-quality wine producers in consolidated vineyards.[30]

Nevertheless, the pace of change is slow. In the whole Community, for example, only 200,000 out of 2.7 million hectares of vines were grubbed-up between 1976 and 1982, and then mainly in France where vines have been disappearing at a rate of approximately 25,000 hectares a year, although in 1980/1 the total rose to 41,000 hectares. When there are nearly half a million hectares of vines in the Languedoc-Rousillon region alone, these figures achieve a clearer perspective. Faced by such 'structural rigidity' in the vine-growing regions, the attention of the Commission has turned more recently to increasing the consumption of wine in countries such as the

United Kingdom and Ireland. Here national rates of excise duty, which appear to discriminate against wine and in favour of competitive beverages such as beer and spirits, have been called into question.

Sugar beet

Sugar beet is a relatively minor crop in the context of the total agricultural production of the EC; it contributes between 2 and 3 per cent of the total output of most countries. The crop, however, provides a significant source of income to farmers in two limited areas of the Community: in the regions of Italy bordering the Adriatic Sea, from Veneto in the north to Basilicata in the south; in a sweep of regions stretching from Ile-de-France and Champagne-Ardenne in France, through southern Belgium to western regions of the Netherlands, but also including the East Anglian region of the United Kingdom. In Belgium and the Netherlands, for example, 17 per cent of holdings grow sugar beet compared with a figure of 6 per cent for the whole Community. Lower levels of regional specialisation prevail in Italy; the crop provides between 1 and 4 per cent of the total output of most regions and is found nationally on only 4 per cent of all holdings. In Emilia-Romagna, however, 8 per cent of agricultural output is provided by sugar beet. Within the EC, the highest regional values occur in Picardie and Champagne-Ardenne in France, the Polder in the Netherlands, East Anglia in England, and the eastern islands of Denmark. Although West Germany has the highest national specialisation in sugar beet, there is not the same degree of regional specialisation in the crop as is found elsewhere in the Community. The Niedersachsen and Nordrhein-Westfalen regions show the highest regional levels of specialisation in the crop. In countries where sugar beet is a minor enterprise, a very localised pattern of production is evident. In Denmark, for example, the crop is localised in the island of Lolland Falster,[31] a situation that has prevailed since 1872.[32]

Only Ireland and Denmark exhibit falling levels of national specialisation in sugar beet production. Elsewhere in the Community, national specialisation has increased in recent years under the twin stimuli of price support and production quotas. At a regional level, specialisation has increased mainly in those areas already associated with the production of sugar beet. In Italy, Emilia-Romagna and the southern regions have increased their specialisation to the greatest extent, while in the northern countries the regions of Zeeland, Noordbrabant and Groningen in the Netherlands, Brabant and Hainault in Belgium and Nord, Champagne and Picardie in north-east France have been similarly affected. Generally, sugar beet is an integral part of an arable rotation of crops in these areas, although yields vary markedly, being highest in France, West Germany and the Netherlands, and lowest in Italy. The scale of production is largest in France and the United Kingdom with an average of 11 and 14 hectares of beet per farm where the crop is grown, whereas in Italy and West Germany the crop is more fragmented: averages of 3.4 and 4.9 hectares per farm respectively. Elsewhere, sugar beet has declined relatively in the output of agricultural regions, especially in central France, the Veneto region of northern Italy, and all parts of Denmark (Table 25). Central to these changes in the regional location of production has been the distribution of the processing factories which allocate the production quotas, together with the incidence of soils favourable to sugar beet production – for example the loessic deposits of the Braunschweig area of West Germany.

Vegetable oils and fats

A distinction can be drawn between the production of vegetable oils from tree and field crops. Olive trees exemplify the first mode of production, and oilseed rape the second. As a Mediterranean crop, the commercial production of olive oil is highly localised within the Community and achieves a regional significance in only two areas of Italy – Calabria and Puglia. In these two regions nearly a quarter of the value of agricultural output is generated by olive oil, and that proportion has been rising in recent years. Elsewhere, olive oil is a relatively small (less than 7 per cent of output) and declining element in agriculture, although olive groves remain a characteristic and traditional element of the rural landscape of southern Italy and parts of southern France. Nevertheless, production is greatly fragmented with an average size of between one and three hectares of olives per holding, while yields are low. Consumption per head of olive oil has been falling since 1974 and, in common with many other products, surpluses are common. Consequently, the profitability of olive oil is largely maintained by direct financial aids to producers under the CAP, and since 1979 by subsidies on consumption. Considerable increases in payments from the Agriculture Fund are expected now that Greece is a member of the Community, while the financial consequence of supporting the product has been a major obstacle to the prospective membership of Spain and Portugal. These countries are also major producers of olive oil. Moreover, since the exact area of olive trees in the Community is still not known, the system of support is open to fraud.[33]

Oilseeds produced from field crops have a contrastingly dispersed pattern of production within the EC. One broad area where the crop has developed an importance lies in central and north-west France from Centre in the north to Midi-Pyrénées in the south. Here colza and rape are included in an arable rotation of crops and can be farmed using mechanised equipment common to cereals. A second broad area encompasses the most southerly regions of Italy, particularly Campania. Oilseed crops such as sunflower, castor and soya tend to characterise production in these areas. Elsewhere, oilseeds have a fragmented pattern of importance: Umbria in central Italy, Alsace in north-east France, Polder in the Netherlands, and Schleswig-Holstein in West Germany.

No clear regional pattern is evident in the changing specialisation in oilseeds. Amounts of increase and decrease are small and randomly distributed, although a majority of regions in most countries show an increasing specialisation in oilseed production (Table 25). More recent data would reveal marked increases in oilseed rape in the United Kingdom, Denmark and Ireland. In the United Kingdom, for example, it has become established as a profitable break-crop in cereals production in the East Midlands region. In common with countries such as Belgium and the Netherlands, oilseeds were not given price support under national agricultural policies prior to the CAP. Consequently, membership of the Community has resulted in a considerable price increase for oilseed rape to which farmers have responded.

Conclusion
This chapter has developed the theme of the increasing regional specialisation of agriculture in the EC. A majority of regions have become more specialised in cereals, oils and fats, wine and non-CAP products, but only in a minority of regions is there a similar trend for sugar beet, fruit and vegetables (Table 25). There is some variation in trends within the Member States. A majority of regions in Denmark, France, West Germany and the United Kingdom, for example, have become more specialised in cereals, whereas falling levels of regional specialisation are common in Italy, Ireland, Belgium and the Netherlands. Nevertheless, the spatial consequences are similar: specialisation has tended either to increase by the greatest amount, or decrease by the least amount, in those regions where a crop contributes most to the total value of agricultural output. In this way, pre-existing variations in the location of agricultural production within the Community has been reinforced.

Such developments are entirely to be expected in the context of a common market. Prior to the EC, regional advantages in production costs tended to exert a mainly national influence. However, with the freer, competitive trading of agricultural products within a common market, such regional advantages have assumed a Community-wide significance. Clearly, the cost-competitiveness of any region can be modified by changes in its farm-size structure and by national factors such as the rate of inflation, currency exchange rate and domestic policy measures, including technical barriers to trade. But to date, researchers have been unable to isolate the regional effects of these various factors. Regional agglomeration economies can further complicate the situation, as cost advantages can be gained by producers in regions with a strong network of co-operatives, research and extension services, food processing industries or credit institutions.

The CAP appears to have modified the process of regional specialisation in two ways. First, the CAP created a low-risk economic environment in which individual farmers have been able to abandon mixed farming and concentrate on fewer production systems. While an increasing regional specialisation is a world-wide phenomenon in agriculture, and was certainly present in Member States prior to the formation of the EC, the process has probably been accelerated by the security offered under the CAP. Not only have prices been held stable and above those ruling on the world market, but Community produce has been guaranteed a preference in the domestic market and has been subject to export refunds on the international market.

The second impact of the CAP has been to modify the process of specialisation according to the degree of market regulation offered to the various agricultural products of a region. Regions specialising in crops subject to price guarantees, intervention and market regulation, for example, have been able to increase their specialisation despite the production of surpluses and largely in the absence of inter-regional competition. On the other hand, products without strong market regulations have been subject to inter-regional competition based on variations in production costs. Regional specialisation in fruit and vegetables, for example, indicates a real movement in the location of agricultural production Thus the conditions under which specialisation is taking place tend to vary with the type of agricultural production in a region, although direct rather than inferential evidence in matters such as regional costs of production remain sparse. There is a clear need for more research of a comparative nature between regions where agriculture is becoming more specialised in a particular product.

Two important ramifications follow from the increasing regional specialisation of production. First, the structural rigidity of production is increased in so far as individual farmers and regions become dependent on fewer agricultural products. With capital, farming skills, and systems of production and marketing committed to two or three products, individuals and whole regions find it increasingly difficult to alter production in line with market demand. The problems of rigidity in the production of products already in surplus are well known in the vine-growing areas of the Midi, as well as the milk-producing regions of many upland areas. Expensive price-support and structural policies have been required to ameliorate the socio-economic problems of such regions. Agricultural trends under the CAP, however, appear to be creating more, rather than fewer regions dependent on a limited number of products, and this can only increase the difficulties of policy making in the future.

The second outcome of regional specialisation lies in the intensification of many of the 'environmental' problems associated with modern farming techniques. These include soil compaction from the use of heavy farm machinery, eutrophication and nitrification of water resources following the washing of fertilisers from the land into drainage systems, the impoverishment of the flora and fauna of the countryside due to the drainage of wetlands, the ploughing of heathland, and the grubbing-up of small woodlands and hedgerows. All of these tendencies are well-documented and appreciated but are intensified when large areas are given over to the specialised production of certain types of crops or livestock. While the CAP cannot be blamed for causing these changes in farming practices, existing policies do nothing to counteract their environmentally damaging consequences. Indeed, the CAP fails to express any environmental concern;[34] rather it still reflects objectives established nearly thirty years ago which are now defended by the politically powerful farming lobby. For as long as wider economic and political considerations dominate policy making in Brussels, any environmental policy for agriculture will emanate piecemeal from individual Members of the Community. In these circumstances, restrictive environmental policies are unlikely, for if they were to be applied in some countries and not others, the competitive economic strength of agriculture in these countries and regions could well be damaged.

Notes

1 These themes are explored by M. Tracy, *Agriculture in Western Europe: challenge and response 1880–1980* (2nd ed.) (London, Granada, 1982).
2 B. Andreae, *Farming, development and space: a world agricultural geography* (New York, de Gruyter, 1981), pp. 197–231.
3 N. J. G. Pounds, 'Agriculture in the Common Market', *The East Lakes Geographer*

I, (1964), pp. 40–52; I. B. Thompson, *Modern France: a social and economic geography* (London, Butterworths, 1970); B. W. Ilbery, *Western Europe: a systematic human geography* (London, Oxford University Press, 1981); R. E. Mellor, *The two Germanies: a modern geography* (London, Harper and Row, 1978); J. T. Coppock, *An agricultural atlas of England and Wales* (London, Faber, 1976); D. A. Gillmor, *Agriculture in the Republic of Ireland* (Budapest, Akademiai Kiado, 1977); J. Tuppen, *The economic geography of France* (London, Croom Helm, 1983); A. H. Kampp, *An agricultural geography of Denmark* (Budapest, Akademiai Kiado, 1975).

4 Most sources give only regional data on crop areas and livestock numbers and for censuses taken in different years in each Member State. *The Yearbook of Regional Statistics* (Eurostat) has been available for The Nine only since 1981 and provides a limited range of agricultural data for 1975 and 1977. Earlier regional data refer to The Six for 1968. Data for 1975 are also reproduced in the *1978 Report* of the Commission's publication: *The agricultural situation in the Community*, pp. 284–306.

5 For example R. Lösch *et al*, *Die landwirtschaft in den regionen der EWG und ihre verbindung zu den anderen wirtschaftsbereichen*, Studien zur Agrarwirtschaft 8, (Munich, Info-Institut für Wirtschaftsforschung, 1971); H. Schmidt, *Die lage der EWG-landwirtschaft in wirtschaftlichen an passungsprozeb* (Munich, BLV Verlagsgesellschaft, 1969).

6 Imputed monetary values (ECUs) have been ascribed to the agricultural output of eight crop and livestock products, with commodities not covered by the CAP forming a ninth category. Measures of value allow agricultural production from a variety of sources to be compared directly. The data for 1976 are problematic in that production trends were partly disrupted by widespread drought conditions in the EC. Consequently only general developments in the regional structure of agricultural production can be inferred. Data have been drawn from Annex 1 and 2 of: P. Henry, *Study of the regional impact of the Common Agricultural Policy*, Regional Policy Series 21, Regional Impact of the C.A.P. Working Group (RICAP), (Commission of the European Communities, Brussels, 1981). In Figures 7 and 9–13, parts (a) are revisions of maps published in various Reports by the Commission; parts (b) are based on data in the RICAP publication.

7 Statistical Office of the European Communities, *Agricultural Statistics 1973*, Eurostat, (Brussels, 1974), pp. 216–57; and I. R. Bowler, 'The CAP and the space-economy of agriculture in the EEC', in R. Lee and P. E. Ogden (eds.), *Economy and society in the EEC* (London, Saxon House, 1976), pp. 235–55.

8 The distinction is discussed by S. Tarditi, 'Analysis of the location of horticultural production', *Acta Horticulturae* XXV (1972), pp. 250–64.

9 Kampp, *An agricultural geography of Denmark* (1970).

10 The regionalisation of cereal prices was also abolished in 1976/7.

11 I. R. Bowler, 'Regional specialisation in the agricultural industry', *Journal of*

Agricultural Economics XXXII, (1981), pp. 43–54.

12 M. Butterwick and E. Neville-Rolfe, *Food, farming and the Common Market,* (London, Oxford University Press, 1968), p. 98.

13 F. Plet, 'Co-opération, productions contractuelles et transformations de l'espace en Limagne', *L'Information Géographique* IIIL (1983), pp. 12–22.

14 L. Tirone, 'La vigne dans l'exploitation agricole en Italie, *Méditerranée* I, 339–62.

15 Withdrawn produce cannot be placed back on the open market. Consequently it is either destroyed (cauliflowers, tomatoes, peaches), fed to livestock (tomatoes, apples), processed (peaches, pears, apples) or distributed free to institutions (table grapes). Relatively small proportions of total production are withdrawn from the market in any year – for example 11 per cent of apples in 1975/6, nineteen per cent of peaches and 17 per cent of oranges in 1976/7, and 15 per cent of mandarins in 1978/9. During the period of study cucumbers, plums and cherries had only a reference price system.
House of Lords Select Committee on the European Communities, *Fruit and vegetables,* 22nd Report, HL 147 (1980/1).

16 The area under fruit rose from 2,228ha. in 1950 to 18,529ha. by 1965. R. C. Rickard, *Regional planning and horticulture in France,* Political and Economic Planning Bulletin 501 (London, 1968), p. 146.

17 I. B. Thompson, *The lower Rhône and Marseille* (Oxford, Oxford University Press, 1975), pp. 23–30.

18 Tarditi, *Acta Horticulturae* XXV, (1972), pp. 250–64.

19 L. Hinton, 'Outlook for horticulture in Europe', *Outlook on Agriculture* IX (1977), pp. 108–13.

20 Centre for European Agricultural Studies, *Survey of the potato industry in the EEC,* Report 8 (Wye College, University of London, 1975).

21 The extensive literature on wine production is reviewed by J. R. Dickenson and J. Salt, 'In vino veritas: an introduction to the geography of wine', *Progress in Human Geography* VI (1982), pp. 159–89.

22 Regulations 816/70 and 817/70, later amended by 337/79 and 338/79. Similar categories have been used in France for some years (but not in Italy): wines with a registered designation of origin (appellation d'origine controlée-AOC), delimited wines of superior quality (vine délimités de qualité supérieure-VDQS), and local wines (vins de pays). In addition a distinction can be drawn between red (the majority) and white wines.

23 Dickenson and Salt, *Progress in Human Geography* VI (1982), pp. 159–89.

24 Greece now forms a fifth centre of wine production in the EC, contributing 4 per cent of total output. Wine is produced by a third of all holdings, the average area of vines is 0.5 hectares per holding, and 92 per cent of output is classified as table wine.

25 European Communities, *Wine in the European Community,* European Documentation 2–3/1983 (Brussels, Office for Official Publications, 1983), p. 20.

26 I. Stevenson, 'Sour grapes for rich harvests', *Geographical Magazine* IIL (1976), pp. 262–4.

27 The regionalisation of regulations for wine was first introduced in 1970 and elaborated by Regulation 337/79 in 1979. Broadly, the five regions (with some sub-divisions) are A: Belgium, Luxembourg, Netherlands, United Kingdom; B: West Germany (Baden-Württemberg) and France (Nord and Pays de la Loire); CI: south-western France, northern Italy; CII: central Italy and southern France; CIII: Corsica, southern Italy, Greece. European Communities, *Wine in the European Community* (1983), p. 42 and 79.

28 P. Carriére, 'Viticulture et espace rural', *Bulletin de la Société Languedocienne de Géographie* VII (1973), pp. 221–37.

29 P. Bartoli and M. Meunier, *La politique de reconversion viticole*, Série Etudes et Recherches 66 (Montpellier, Institut National de la Recherche Agronomique, 1982), p. 61.

30 C. Roux, 'Heures et malheurs des coteaux rhodaniens dans la région valentinoise', *Revue du Géographie de Lyon* LIII (1978), pp. 23–35.

31 M. Cabouret, 'Quelques traits de l'évolution récente de l'économie agricole danoise', *Annales de Géographie* CXXXIII (1974), pp. 684–714.

32 Kampp, *An agricultural geography of Denmark* (1970), p. 32.

33 House of Lords Select Committee on Agriculture, *Olive oil*, 3rd Report, HL 31 (1982/3).

34 European Environmental Bureau, *The Common Agricultural Policy and its impact upon nature and the environment* (Brussels, EEB, 1978), p. 32.

8
Regional trends in the specialisation of agriculture: livestock products

Attention is now turned from crops to livestock products which tend in aggregate to dominate agricultural production in the Community. Here, the main objective is to determine if the conclusions drawn for crops, on the greater regionalisation of agriculture, can be generalised to other products.

Milk and dairy products
Milk is the main problem commodity in the EC. Each year between 10 and 15 per cent of production cannot be sold on the normal market. This 'structural' surplus absorbs a major share of the expenditure of the Agriculture Fund through the support of prices for dairy products on the internal market and the disposal of surplus production on the world market. For social reasons, the prices of dairy products can be reduced only slowly in real terms: a large number of farmers (nearly two million) depend wholly or mainly on milk for their income; milk producers tend to occupy the smaller farms in the Community on which opportunities for alternative forms of production are severely limited. Yet the importance of dairy farming varies considerably within the EC and changes have been taking place in the structure and location of milk production under the CAP.

Over the last two decades, countries of the Community have been on diverging paths so far as the contribution of milk to the total value of agricultural output is concerned (Figure 8). Luxembourg, Ireland, and the Netherlands, for example, have become increasingly specialised in milk production, whereas the opposite trend is evident for France, Belgium and, to a lesser extent, West Germany. Nevertheless, a majority of countries, including Italy, now have a greater share of total output from milk compared with two decades ago. Countries can be divided into two groups as far as the structure of dairy farming is concerned. In the first group are countries with relatively large-scale, technically-sophisticated dairy farms (United Kingdom, the Netherlands, Denmark, Luxembourg). In these countries the average dairy herd is in excess of twenty cows, a high proportion of herds are larger than forty cows in size, milk yield per dairy cow and output per dairy farmer are high, and the proportion of farmers wholly or mainly dependent on the production of milk for their income is above the average for the Community (Table 30). The six remaining countries comprise the second group and display converse characteristics: dairying remains essentially an activity associated with small farms; dairy cows tend to play a relatively minor role on the farms where they are kept even though the proportion of farms with dairy cows can be relatively high – 60 per cent in the case of West Germany. Ireland is to a degree exceptional within the second group of countries in having a relatively high proportion of dairy farms which are nevertheless small in herd-size, milk yield per cow and output per dairy farmer.

At the regional level, the location of dairying is influenced by the following factors:[1] a high rainfall and long growing season which gives an advantage to grass compared with crop production; a small size-structure of farming under which an intensive form of agriculture is needed for economic viability; proximity to market which has been an historical advantage from times when transport costs for liquid milk were high. These features occur either alone or in combination in three broad regions where dairying is an important element of agriculture:[2] 'coastal' locations including the mid-west and south-west of Ireland, the Gelderland, Drenthe and Utrecht provinces of the Netherlands, northern areas of Belgium, peripheral areas of Wales and south-west of England, Brittany and Normandy; 'continental mountain' locations including Bayern (Bavaria) and Baden-Württemberg, Lorraine, Franch Comté (northern Alps) and the Auvergne; 'north Italian' locations including Valle d'Aosta, Piemonte, Veneto and Lombardy (Figure 11a). The density of dairy cows is highest in the Netherlands, northern Belgium, Brittany and Lombardy, while the use to which milk is put varies throughout the Community. Liquid sales, for example, are highest in the United Kingdom and Italy (Table 30), whereas butter (southern Ireland), cheese (Auvergne), cream and even condensed milk (Normandy) characterise other regions. Processing and distribution in most countries is in the hands of producer co-operatives.[3] In the Netherlands and West Germany about three quarters, and in Belgium nearly half of the creameries are run co-operatively, while in the United Kingdom the Milk Marketing Board operates its monopoly purchasing powers as a type of producer co-operative. The number of private and co-operative creameries continues to fall as the process of concentration into fewer, larger units takes hold of the milk processing industry.

A variety of feeding systems operate within the broad regional groupings just noted.[4] Dairying in Ireland, for example, is traditionally based on summer milk production from spring-calved cows; Brittany has a climate well-suited to the production of maize-silage and this is used for dairy cows as part of a forage-dominated diet. In the Netherlands, a country with only moderate potential for forage growth but with close proximity to North Sea ports, high concentrate-feeding and intensive stocking rates are combined to produce a high output of milk per hectare.

Under the CAP, regional variations in the production of milk have been drawn more clearly. An increasing regional specialisation in dairying has been most evident within Ireland and Denmark (Table 25). In Ireland, south and south-western districts have become more dependent on milk production as have western areas of Jutland.[5] In Belgium and France, by contrast, regional specialisation in milk has fallen by the greatest extent. In France, for example, milk has contributed less to the total output of a broad sweep of regions in the north, centre and east of the country. Only in the north west (Brittany, Normandy and Maine) and the Massif Central has dairying become more significant in agricultural production (Figure 11B). Most regions of Italy have developed milk production, especially Trentino-Veneto in the north and Lazio-Molise in the centre; but given the limited scale of development of dairying the trend should be interpreted as diversification rather than specialisation in the structure of agricultural production.

These developments in the location of production have taken place within the context of structural changes in dairy farming (Table 31). Where profitable alternatives have existed, farmers have left dairy production. Over the last decade, for example, 1.5 million farmers stopped producing milk at an average rate of 4.5 per cent a year. Consequently, the number of dairy cows has also been falling, albeit at the low annual rate of 0.2 per cent. In countries such as France and Belgium, many milk producers have been able to switch to cereals, beef and intensive livestock (pigs, poultry). The smallest farms have shown the greatest rate of decline and production has become more concentrated in herds of over thirty cows. But milk yields per cow have continued to rise – in Italy by 34 per cent between 1970 and 1979 (Table 31) – and the total volume of milk production has increased in every Member State. The following developments have been responsible: the greater use of high-yielding Freisian cows and improved breeding through artificial

Table 30. The structure of dairy production

Country	% holdings with dairy cows	Dairy farms as % total farms^a	% cows in herds over 40 cows	% herds over 40 cows	Ave. number cows/herd	% production in liquid sales
West Germany	60	18	11	3	12	10.4
France	46	23	17	4	14	9.9
Italy	19	4	25	2	6	24.8
Netherlands	54	38	62	31	32	7.3
Belgium	52	14	23	7	17	18.5
Luxembourg	64	20	29	12	21	n.d.
United Kingdom	27	19	83	52	53	48.6
Ireland	53	25	35	9	14	9.6
Denmark	44	14	39	16	23	6.5
EC	35	13.5	32	7	14	n.d.
(Greece)	(14)	(n.d.)	(n.d.)	(n.d.)	(2.8)	(n.d.)

a: 1975 Farm Structure Survey

Source: Commission of the European Communities, *Agricultural Situation in the Community. 1981 Report*, (Brussels, 1982), pp. 270–2, 291–9; *1982 Report* (1983), pp. 319, 320, 322

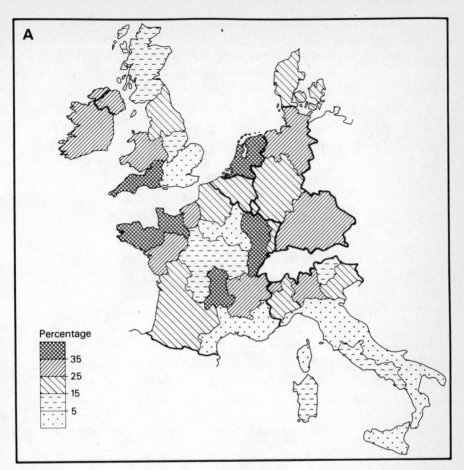

Figure 11A Milk as a percentage of the total standard gross margin of the region, 1975

insemination; the increased adoption of intensive methods of feeding concentrates; improvements in the age-structure of dairy herds (fewer, older cows); the introduction of machine-milking techniques which have allowed larger herds to be handled by an individual farmer.

These yield-increasing technical improvements in milk production have largely thwarted attempts under the CAP to accelerate structural changes in the dairy sector. By the end of 1980, for example, the Dairy Herd Conversion Scheme (1977) had taken approximately 1.5 million cows out of production, but the volume of milk involved was more than offset by rising yields from the remaining cows. Similarly the co-responsibility levy (1977) appears to have

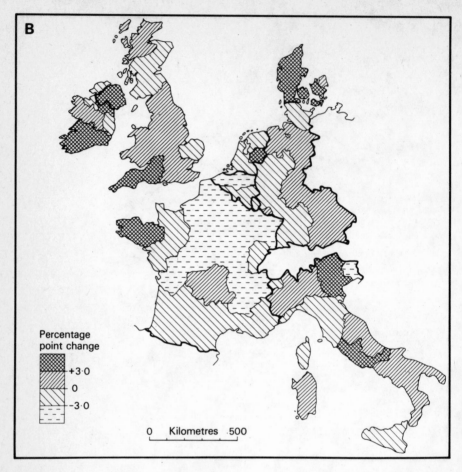

Figure 11B. Changing percentage of milk in the total agricultural production of the region, 1964/5–1976/7

worked against the interests of large 'efficient' producers[6] acting more as a method of increasing the revenue available for disposing of surplus production (156 million EUA in 1978) than as a disincentive to increased output.

Attempts to determine if dairying under the CAP has become located in countries with more 'efficient' production have proved inconclusive.[7] On the basis of costs per kilogramme of milk (1980), France, Belgium and Ireland emerge as the lowest-cost producers, with West Germany, Italy and the Netherlands as the highest. However, technical advantages in milk

Table 31. Changes in the dairy sector, 1970–9

Country	% change in number of dairy cows	% change in milk yield/ cow	% change in milk deliveries to dairies	Milk yield per AWU[a] (tonnes)[b]	Milk yield per cow (kg)[b]
West Germany	−2.2	+16	+20	113	4550
France	0	+14	+27	89	3665
Italy	−4.4	+34	+14	56	3362
Netherlands	+23.4	+16	+45	198	5035
Belgium	−1.9	+7	+15	125	3855
Luxembourg	+8.0	+10	+26	n.d.	3975
United Kingdom	+0.3	+21	+26	190	4880
Ireland	+33.7	+30	+66	96	3234
Denmark	−6.3	+25	+17	220	4855

a: AWU – annual work unit; b: 1980
Source: Milk Marketing Board, *Dairy facts and figures 1979*, (London, 1980); Food and Drink Industries Council, *The food industry and the farmer*, Occasional Paper (London, 1982), pp. 4–5

production have tended to be overshadowed in the 'competitiveness' of national dairy sectors by economic advantages gained through prevailing exchange rates and national inflation rates. During the 1970s, for example, the ratio of milk price to production costs declined to the greatest extent in the United Kingdom and the Netherlands, fell by the least amount in Ireland and Belgium, and actually rose in Italy to the advantage of dairy farmers. In addition, problems are posed by the inclusion or exclusion of interest charges in calculations of 'competitiveness', as well as the cost of family labour. The United Kingdom, for instance, is more dependent on hired labour (44 per cent of total labour inputs) than other countries (all less than 19 per cent). Only when imputed family labour costs are added is the technical superiority of British dairy farming clearly revealed. At the regional level, areas with a prior specialisation in milk production have increased that specialisation under the CAP. This development, however, is not related to the level and trend in regional milk prices but more to regional ratios of land (farm size), labour and capital and the existence of alternatives to dairying.[8]

Nevertheless, the outlook for milk production in the EC is not good. The application of existing dairy technology (breeding and feeding) throughout the Community could raise the output of milk by at least 2.5 per cent a year. The increased production would probably come from those countries with a low level of productivity at present (France and Ireland); diminishing returns in the use of expensive purchased feeds seem to have been reached in countries such as the Netherlands and the United Kingdom.

Paradoxically, any downward revision of the real price of milk is likely to be felt more severely by large-scale 'efficient' producers than by their small-scale counterparts.[9] Large-scale producers, especially those in the United Kingdom, are particularly dependent on a high-cost technical performance and are unable to take a cut in the 'family' wage as is open to smaller producers.[10] Moreover, smaller farms are less likely to go out of production since there are few profitable alternatives.[11] Thus, with a price reduction of 15 per cent, the greatest losses of production would occur in the United Kingdom and West Germany, and the lowest in France and Italy. A price reduction on this scale, however, would be necessary to bring supply into equilibrium with market demand.[12] Conversely, a selective programme of aid to compensate small producers leaving milk production would have a greater impact in France, Ireland and Italy. Once again, national interests are at stake in the development and application of policy measures. The strong corrective measures of 1984, to counteract over-supply, are meeting with resistance in many countries, while Ireland has been excused restrictive quotas on production.

Beef and veal
In Western Europe, beef comes mainly from dairy or dual-purpose cows slaughtered at the end of their milking lives, or else from calves from the dairy herd. Such calves are generally cross-bred from a beef bull and a cow of a dairy breed. Thus 80 per cent of holdings with cattle depend for their income on a combination of milk and beef production: the monthly income from milk helps to finance the one to two year gestation period for beef production. Specialist beef breeding cows are a feature of agriculture mainly in Ireland and the United Kingdom, and even in these countries there is a trend towards more production from dairy herds: beef breeding herds tend to be economically viable only in regions where supplementary production subsidies are paid under the Less Favoured Areas Directive. The pure beef breeds such as Charolais and Limousin in France, Romagnola, Marchigiana and Chianina in Italy, and Hereford and Aberdeen Angus in the United Kingdom account for a minor proportion of beef production in each country.

Nevertheless, beef is produced in a variety of ways. For example, the cross-bred calf can be reared on whole or skim milk with a little roughage, and kept closely confined in a pen before slaughter at 102–229 kilogrammes when a few weeks old (vealers). On the other hand, young cattle can be fed intensively on cereals to produce 'baby beef' at a higher slaughter weight at fourteen to eighteen months old. With the adverse consumer reaction to the use of hormone growth promoters in veal, 'baby beef' has developed in countries such as Denmark, Belgium and West Germany. Alternatively, the young cattle can be single-suckled from a breeding cow and fattened in yards on cereals and fodder crops to a slaughter weight of 356–406 kilogrammes after twelve to fourteen months. Older male fat cattle (eighteen to twenty months) are often finished on maize silage, grass silage and pasture at weights of up to 508 kilogrammes, while mature cattle (two to four years) can be fattened from grass and by feeding in open yards to slaughter weights of 508–711 kilogrammes.

The type of production system selected by a farmer tends to vary according to farm size.[13] On holdings of less than thirty hectares, for example, dairying tends to be the dominant enterprise with calves and cull-cows the main beef product. On larger farms the fattening of stores from cereals is possible, while on the largest farms grass fed cattle can be produced. Large feed-lots which fatten purchased store cattle, however, remain an exception and are found mainly in areas such as the Paris basin and the Po valley of northern Italy. Many of the feed-lots in France are owned co-operatively and fatten the young store animals produced by members of the group. Profit margins vary from year to year and with the system of production. In most years in the United Kingdom, for example, the order of profitability per hectare is eighteen-month beef (highest), yard-finished grazed stores, lowland suckler-herds and grass-finished stores.

The flexibility of beef production, and its relatively low capital cost, have resulted in a widespread distribution of cattle within the EC. Until recently, the agricultural statistics of the Community did not distinguish between dairy cows and beef cattle, but now it appears that approximately half of all holdings contain beef cattle with proportions varying from 90 per cent in Ireland to nearly 30 per cent in Italy. (Table 32). Because beef production tends to be a minor enterprise on most farms, only 4 per cent of all holdings in the Community can be classified as cattle rearing/feeding farms. In Ireland, however, nearly 30 per cent of farms can be classified as specialising in beef production, and in the United Kingdom nearly 20 per cent.

The regional distribution of beef production has strong parallels with dairying since the two types of farming are functionally related. Thus the three broad regions of dairy farming can also be discerned in the locational pattern of beef production (Figure 12A). The two main discrepancies between the locations of milk and beef production occur in Scotland and the

Table 32. The structure of beef production

Country	% holdings with cattle	Cattle Rearing and Feeding farms as % total farms[a, b]	% cattle in herds over 100 cattle	% herds over 100 cattle	Ave. number cattle/herd
West Germany	69	1.1	13	3	27
France	63	6.1	19	4	33
Italy	29	0.8	28	1	31
Netherlands	64	3.2	40	14	57
Belgium	70	5.8	22	6	37
Luxembourg	75	3.5	40	16	56
United Kingdom	70	19.0	64	25	75
Ireland	90	29.1	24	5	32
Denmark	58	0.5	37	12	46
EC	50	4.2	29	5	31
(Greece)	(19)	(–)	(8)	(0)	(8)

a: 1975 Farm Structure Survey; b: excludes Dairy with Beef, and Mixed Dairy/
Beef farms (EC = 10.6 per cent of all farms)
Source: As Table 30

Netherlands. In the former, specialist beef production occurs in the absence
of dairying, especially in north-east Scotland; while in the latter,
specialisation in dairying has not led to the widespread development of
ancillary beef production. There has, however, been some development of
veal production for export from the Netherlands to Italy. Regions can also be
grouped according to their dominant system of production. Specialist beef
herds, for example, characterise production in Scotland, northern England,
Northern Ireland, Ireland, and west-central France (Massif Central,
Limousin); fattening on dairy farms is a feature of Belgium, south-west
England, Wales, north-west France and Denmark; fattening purchased
calves with fodder crops is associated with northern Italy, Bavaria and
Württemberg.

The structure of beef production reflects the farm size distribution in each
Member State (Table 32). Herd sizes are largest in the United Kingdom, for
example, with a majority of cattle in herds of over one hundred animals.
Many herds are comprised of specialist beef stock. Relatively large herds also
characterise production in the Netherlands and Luxembourg. In Ireland, by
comparison, herd sizes are small despite the importance of beef production,
and similar herd sizes are recorded in France, West Germany and Italy.
Because of the subsidiary nature of the beef enterprise on many farms,
producers can move into and out of, or expand and contract production

according to the profitability of beef production. This has created a production cycle in the output of beef. The CAP appears not only to have brought the various national cycles more closely into phase, but to have made their extremes more frequent.

Production trends have done little to alter the relative national specialisation in beef production in the last decade. Only Luxembourg and Italy (increase), and Ireland (decrease), have experienced significant changes, and 35 per cent of agricultural output is still accounted for by beef in the latter country (Figure 8). Nevertheless, quite substantial regional

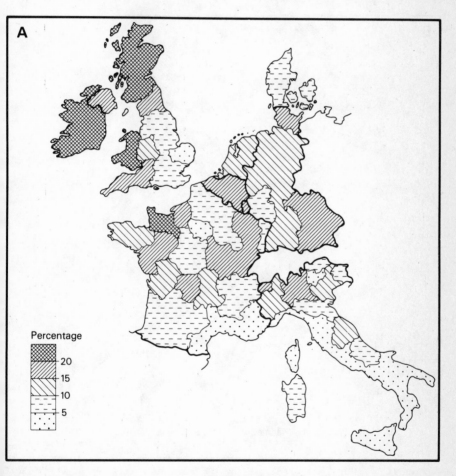

Figure 12A. Beef and veal as a percentage of the total standard gross margin of the region, 1975

changes in specialisation have occurred within several countries (Table 25). An increasing regional specialisation in beef has occurred throughout Ireland, in southern parts of Belgium including Luxembourg, and in northern and south-eastern regions of West Germany. At the other extreme, falling specialisation has characterised a majority of regions in the Netherlands, especially in eastern areas, and Denmark. In the remaining countries, those regions with an important beef sector have tended to become more specialised in the product, while in Italy there has been a widespread but not intense development of beef production (Figure 12B).

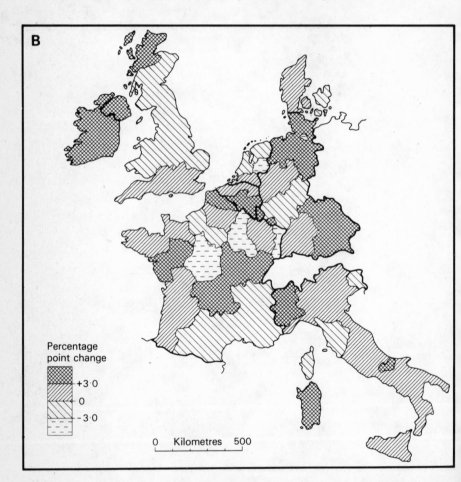

Figure 12B. Changing percentage of beef and veal in the total agricultural production of the region, 1964/5–1976/7

The financial assistance under the CAP for rearing beef calves (Italy and Ireland only) and keeping breeding cows (Suckler Cow Premium)[14] has furthered the development of beef production in recent years. In addition, many of the programmes designed to stem the output of milk have channelled farmers towards beef production (see Chapter 5). The Hill Livestock Compensatory Allowances (HLCAs) under the Less Favoured Areas Directive (Figure 4) have also encouraged grass-based breeding herds. In the United Kingdom, for example, the Allowances are paid in respect of the number of breeding cows in a regular breeding herd (£44.50 a head in 1982). The Allowances have served to continue the concentration of beef breeding in upland areas, a trend already evident under the stimulus of the previous national Hill Cow Subsidy.[15] The proportion of all breeding cows located in hill areas rose from 57 per cent in 1970 to 63 per cent in 1982, against a trend of declining numbers of breeding cows in the United Kingdom. Nevertheless, most of the available evidence points to producer prices being more influential in decision making on beef production[16] although the substantial price rises under the CAP, both in the 1960s (23 per cent 1964–8) and the 1970s (41 per cent 1971–4), have not been sufficient to make beef a viable proposition as the sole enterprise on most farms. Indeed, the development of beef production in recent years has taken place mainly on larger holdings with more than sixty beef cattle.

Pigs, poultry and eggs

Intensive livestock production includes pigmeat, poultrymeat and eggs. Each is characterised by an increasing concentration into very large, capital-intensive, factory-like production units largely divorced from agricultural land. Animal feed is purchased rather than grown, for example, and the livestock remain inside buildings during their productive lives. Interestingly, the feed processing industry is itself highly concentrated. In France, for example, twenty-seven factories control half the national production of animal feed, while nearly a third of production is located in Brittany.[17]

Two groups of countries can be identified in the structure of intensive livestock production. In the Netherlands, the United Kingdom, Denmark and Belgium the production of pigmeat, poultrymeat and eggs is highly concentrated. Average herd and flock sizes are large, the proportion of holdings with each type of livestock tends to be below the average for the EC, but relatively large numbers of holdings specialise in the production of each type of livestock (Table 33). In the Netherlands, for example, nearly 5 per cent of holdings specialise in pig production, the average broiler flock is nearly 18,000 birds, and only 6 per cent of holdings have laying hens.

In the second group of countries – West Germany, France, Italy, Luxembourg and Ireland – intensive livestock remain essentially a subsidiary farmyard enterprise. Herd and flock sizes are small, a high proportion of holdings farm pigs and poultry, and only a few specialist holdings occur. Thus in West Germany fewer than 2 per cent of holdings specialise in pig production, the average broiler flock is only 180 birds, and 55 per cent of holdings have laying hens.

As with the products previously discussed, the degree of national specialisation in intensive livestock bears little relationship to the structure of production. Turning to pigmeat, a small and generally declining share of the total value of agricultural output is evident in Luxembourg, the United Kingdom, France, Ireland and Italy. Significant shares of output are recorded only in the Netherlands, Belgium, Denmark and West Germany, although declining shares are recorded in the latter two countries (Figure 8). Changes in the number of breeding sows reflect these trends[18] and confirm the second group of countries as the focus of pigmeat production in the EC. In Denmark, for example, 31 per cent of all livestock units are accounted for by pigs.

At a regional level, the specialisation in pig production in northern parts of the Community is even more evident (Figure 13A). These regions have a double locational advantage: they are adjacent to major urban areas where pigmeat products are consumed; they are in areas highly accessible to livestock feed, either produced domestically or imported. Northern locations thus enjoy advantages in the costs of inputs (feed) and marketing over other regions. Since about two-thirds of the cost of production can be attributed to feedstuffs, any savings in transport costs have a significant impact on levels of profitability. Other factors include the efficiency of feed conversion ratios (feed into meat) and the number of weaners per litter. Advisory services in the regions of specialised pig production ensure that the most recent technological developments are applied to these features of production. Consequently, the competitiveness of the leading specialised regions, such as the central and southern provinces of the Netherlands (Noordbrabant, Limburg), is maintained. The processing industries, both of the livestock feed and the pigmeat product, have also gravitated towards these northern regions and thus enhance the agglomeration or external economies of products.

Outlying regions which also specialise in pigmeat can be identified in

Table 33. The structure of intensive livestock production

Country	% holdings with laying hens	Ave. number hens/flock	% holding with broilers[a]	Ave. number broilers/flock	Pigs & poultry as % total farms[b, c]	% holdings with pigs	Pig farms as % total farms	% herds over 400 pigs	Ave. number pigs/herd
West Germany	55	121	12	181	3.0	67	1.4	1.4	41
France	74	64	43	107	1.8	35	0.4	1.8	30
Italy	43	39	31	85	1.1	31	0.5	0.3	9
Netherlands	6	3441	1.4	17,686	8.6	33	4.7	15	205
Belgium	36	347	5	1446	7.8	41	4.6	6.7	116
Luxembourg	56	42	21	16	2.1	40	0.2	0	45
United Kingdom	29	916	1.6	12,755	4.0	15	2.2	14	225
Ireland	47	43	4.7	373	1.1	9	0.2	5.1	114
Denmark	27	186	2.7	2470	1.1	62	0.6	7.4	127
EC	50	103	26	178	2.1	37	0.9	1.7	35
(Greece)	(77)	(23)	(15)	(-)	(-)	(10)	(-)	(0)	(10)

a: poultrymeat; b: 1975 Farm Structure Survey; c: including Mixed Pig/Poultry farms
Source: Commission of the European Communities, *The Agricultural Situation in the Community, 1982 Report* (Brussels, 1983), pp. 319–24

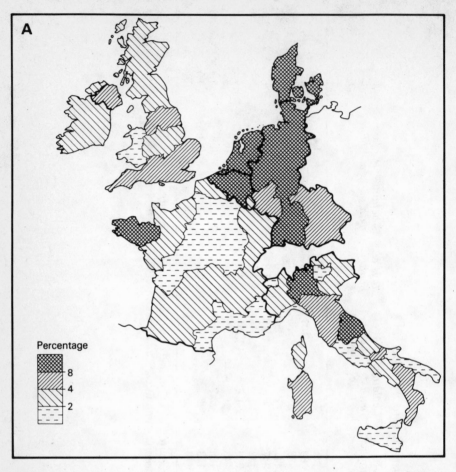

Figure 13A. Pigs as a percentage of the total standard gross margin of the region, 1975

Figure 13A: Brittany, Lombardy and Marche-Umbria. These areas are associated with small farms where an intensive form of agriculture is required for economic viability. The regions of specialised production in Italy tend to serve the industrial cities of the north and they have been supported under the CAP by subsidies on feed imports.

Changes in the location of pig farming have tended to emphasise the existing pattern of production. The number of pigs has increased greatly in Belgium, the Netherlands and northern Italy since the early 1960s, for example, as well as in Brittany and Nord-Pas-de-Calais in France. However,

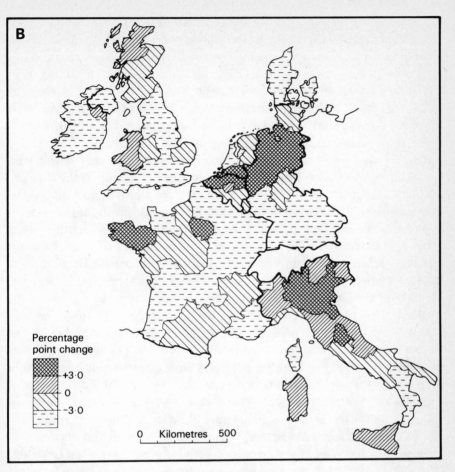

Figure 13B. Changing percentage of pigs/poultry/eggs in the total agricultural production of the region, 1964/5–1976/7

a cyclical pattern has continued to characterise pig production because, like beef, it remains a subsidiary enterprise on many holdings throughout the Community. Low profit margins and production characterised the years 1973, 1976 and 1978, for example, whereas contrary conditions were experienced in 1972, 1974 and 1977.

Poultrymeat contributes a small proportion to the total value of output in most countries of the EC, although the product is more significant in Italy, the United Kingdom, France and the Netherlands than in the remaining countries. Belgium, however, exhibits a varying trend. The output of broilers trebled between 1959 and 1965, production coming mainly from small family farms producing under contract to poultry-feed firms for export. More recently, production has become concentrated into a few large units and the total value of output has declined (Figure 8). Only 5 per cent of holdings now have broilers with an average flock size of nearly 1,500 birds. The regional pattern of specialisation in poultrymeat tends to follow that of pigmeat with concentrations in northern Italy (Emilia-Romagna and Marche), central parts of the Netherlands, northern Belgium and Brittany.

The relative importance of eggs has declined uniformly within the Community, national values tending to converge on the range from 2 to 5 per cent of total output. Only the United Kingdom retains a significant level of egg production with 5.6 per cent of total national output from this source. Regional specialisation in egg production is greatest in south-eastern districts of the Netherlands (Gelderland, Noordbrabant, Limburg), Brittany and the Paris basin within France, northern Italy, northern Belgium, and north-western areas of West Germany. In these regions are also located the majority of the large egg-packing stations and poultry slaughterhouses.[19]

Changes in the regional specialisation of intensive livestock are less complex than for other products (Figure 13B), despite variability in the degree of market support provided by the CAP. While pigmeat is subject to intervention, poultrymeat and eggs receive more external protection (Table 3) than internal price support with the result that considerable inter-regional competition has taken place. When taken together, however, intensive livestock have decreased in regional specialisation within every country of the EC except Belgium and Italy (Table 25). In detail, production has become more focussed in two areas: northern Italy, northern West Germany and the Belgium–Netherlands border. Brittany and the Paris basin form two outlying areas of increasing specialisation. Once again areas with a previous specialisation in a product have had that specialisation enhanced by the forces of economic advantage under the CAP. There are some regions, however, where pigs and poultry are declining from a previous position of importance in the regional agricultural economy. These regions include the east of England, Denmark, and the south-west of West Germany. The level of accessibility to ports for the import of relatively cheap animal feed appears to be a major factor in these locational changes.

Sheepmeat

Sheep are found throughout the EC, but play a prominent role in the national agricultural sectors of only Ireland and the United Kingdom. The per capita consumption of sheepmeat is higher in these countries than elsewhere in the Community, while a significant proportion of holdings farm sheep and can be classified as specialising in their production (Table 34). Also flock sizes tend to be relatively large, especially in the United Kingdom, and sheep account for an important proportion of the total number of livestock. Even so, sheepmeat provides only 4 per cent of the total national agricultural output of each country. Elsewhere in the Community flock sizes are small and sheep form a subsidiary enterprise on a relatively small number of farms.

There are, however, specific regions within France and Italy where sheep form an integral part of the agricultural economy. In France, for example, sheep are localised in three regions, all south of the Loire valley: the north-west fringe of the Massif Central (Limousin-Charentes), Midi-Pyrénées in south-west France, and the southern Alps. An increasing localisation of sheep has occurred in these regions since the early 1960s when cereals began to dominate agriculture in northern France and so displace sheep from *départements* such as Deux-Sèvres and Maine-et-Loire. Three types of sheep farm can be distinguished: small farms, often in mountain areas, with a variety of livestock types and limited amounts of capital (accounting for about a quarter of French sheep production); more specialised family farms where a flock of sheep is often a secondary enterprise helping to expand the farm income without high capital commitment; very large farms, often with hired labour, where sheep are a marginal enterprise. Systems of production vary both by region and size of farm. In the dry hilly areas of south-east France, for example, sheep are the only livestock on farms otherwise specialising in crop production, as the absence of forage limits the development of sheep farming. In grassland areas, such as Nivernais-Charollais in central France, flocks of fifty to sixty ewes have to compete with dairy cows for available forage. While in upland areas, such as the Massif Central, intensive fattening of lambs takes place indoors, often for the winter market. Here sheep are a traditional accompaniment to the production of milk and veal. In arable areas sheep tend to be kept on the least productive land and consume crop residues. Meat is the main product from the flocks, for wool, being viewed as an industrial product, is not covered by CAP regulations. Sheep's milk, however, is increasing in importance especially for the production of cheeses such as the traditional Roquefort type.

In the United Kingdom, by comparison, the sheep sector is regionally integrated. Pure-bred flocks in the hills and mountains of Wales and Scotland produce ewes and lambs for cross-breeding and fattening on lower valley slopes. These cross-bred flocks in their turn produce lambs for

Table 34. The structure of sheep production

Country	% holdings with sheep	Ave. number sheep/flock	Cattle with Sheep farms as % total farms[a]	Sheep as % total nat. livestock units	Ave. annual % growth rate in sheep 1960–70	1970–7
West Germany	4	24	3.4	1.6	−2.0	+4.3
France	14	63	7.9	9.5	+1.1	+1.5
Italy	8	31	3.6	17.3	−0.3	+1.3
Netherlands	14	38	3.0	1.9	+3.6	+4.9
Belgium	8	11	7.3	0.6	+0.7	+3.8
Luxembourg	4	15	1.9			
United Kingdom	30	344	14.5	26.8	−0.1	+0.5
Ireland	21	75	10.3	10.9	+1.3	−1.6
Denmark	3	14	1.4	0.3	+4.2	−2.2
EC	10	87	5.3	11.2	+0.2	+0.9
(Greece)	(22)	(41)	(–)	(–)	(–)	(–)

a: 1975 Farm Structure Survey
Source: Commission of the European Communities, *The Agricultural Situation in the Community, 1982 Report*, (Brussels, 1983), pp. 321–4

fattening on lowland pastures in eastern parts of England and Scotland.[20] There is a multiplicity of local breeds and crosses reconciling the conflicting needs of fleece and meat production. Pure-bred Welsh Mountain ewes, for example, are crossed with Border-Leicester rams, and their progeny are in turn crossed with lowland Suffolk rams in counties such as Leicestershire and Northamptonshire.

Sheep have enjoyed fluctuating fortunes in the EC. Prior to the introduction of CAP regulations for sheepmeat in 1980, their numbers had been increasing to the greatest extent in the Netherlands and Belgium (Table 34) and by the least in Ireland and the United Kingdom. The new regulations, however, with their market support, have proved a general stimulus to production in the EC. In the United Kingdom, for example, a variable premium on lambs and an annual premium on breeding ewes has raised gross margins by as much as twenty-five per cent. Some adjustment in marketing emphasis has been necessary: leaner lambs have been rewarded under CAP regulations while very small lambs have been given a lower rate of premium to the disadvantage of hill farmers. Consequently, hill flocks in Wales and Scotland have increasingly been fed concentrates with the dual objectives of raising the size of lamb produced and fattening some of these lambs rather than leaving them for finishing in lowland areas. Since sheep farming in the hills is also assisted by payments of HLCAs, the localisation of production in upland areas of the United Kingdom has been given further encouragement.

Agricultural regions

Taken in a world context, agriculture in the EC is characterised by a dependence on livestock and by the degree to which capital has been substituted for land and labour. By comparison with other countries, production can be described as intensive: the input of capital and labour per hectare of agricultural land is high as is the output of produce. The presence of a large, affluent, urban population generally accounts for these features, rather than any particular properties of the physical environment.[21]

When the EC is examined in more detail, however, marked regional variations emerge in the character and intensity of agriculture. Some writers claim to have identified a declining intensity of output per hectare away from the focus of urban development within the Community which is located in the London–Amsterdam–Paris triangle.[22] Such a pattern would certainly accord with agricultural location theory predicated as it is on the impact of transport costs, for both the cost of inputs and the price of farm products, with increasing distance from the main urban markets.[23] But more recent research, using the EC Farm Accountancy Data Network, has failed to find any spatial regularity in agriculture that can be attributed to distance from the main conurbations (a core-periphery model). Nor has 'climatic fertility' been identified as an explanatory factor.[24] The complexity of the physical environment and of agriculture itself defies analysis by these general measures, while other factors, such as farm structure and the effect of prior national agricultural policies, also need to be considered. Thus the increasing output of agriculture under the CAP has a complex regional pattern (Figure 7B) comprised of the production trends of a wide range of products.

Nevertheless, an analysis which concentrates on individual products misses one essential feature of agriculture, namely the joint production of many commodities. A number of attempts have been made to summarise the combination of crops and livestock in the regions of the Community, with broadly similar results. Using climatic divisions[25] four regional types of agriculture can be identified: north-west areas with grassland and livestock (cattle and sheep); interior lowland areas based on field crops and cereals; interior dissected upland and mountain zones with livestock; southern coastal fringes characterised by perennial tree crops and vegetables. Sub-divisions of these four agricultural regions can be made[26] using combinations of various crops and livestock from a set comprising cattle/sheep, intensive livestock, extensive crops and intensive crops. Under this method, for example, central parts of West Germany are characterised by cattle/sheep with intensive livestock production. Broadly similar results are obtained when sophisticated techniques, such as factor analysis, are used on a wide range of crop and livestock data.[27] More detail is revealed by these methods such as the subdivision of the grassland systems of livestock production between cattle (Ireland, Normandy-Pays de la Loire, east and central France, upland and mountain areas of northern Italy, southern West Germany), cattle with sheep (the Netherlands) and sheep (Sardinia). Unfortunately, the distinction between milk and beef production cannot be made from the available data. Regions based on field crops, often in association with pigs, are found to characterise eastern England, Denmark, north-central and south-west France, and northern West Germany; while permanent or intensive crop farming dominates agriculture in Mediterranean France, Corsica, Sicily, and coastal areas of north-west and southern Italy. Scotland, Wales, central England and Belgium have more mixed forms of agriculture although sheep and cattle are the predominant types of livestock.

Nevertheless, the degree of specialisation in particular crops or livestock varies between these regions even though products are farmed in combination. Four nodes of specialised agricultural production occur each based on different agricultural products. One specialised region is found in the south of France (Roussillon, Languedoc, Côte d'Azur) and the neighbouring region of Liguria in Italy. Here wine and fruit/vegetables respectively form the focus of specialisation. The second node occurs in central and north-western France and around a Brittany/Ile de France/Champagne–Ardenne axis. Poultry, milk and cereals form the foci of specialisation in succession from west to east. A third node of specialisation, this time in pigs/poultry and non-CAP products, lies in northern Belgium and the Netherlands. Ireland, together with south-western parts of England, Wales and the Highlands of Scotland form the fourth node of specialisation. Milk and beef are the two enterprises that dominate agricultural production in these regions. Elsewhere in the EC agriculture is more diversified, and the four nodes of specialisation, based broadly on either wine, cereals, pigs/poultry, beef or milk, stand out as major elements in the geography of agricultural production.

Nor are these patterns static. Of the census regions of the Community, 67 per cent have become more specialised in the balance of their agricultural production over the last two decades. The minority of regions that have become more diversified are widely dispersed but include the Highlands of Scotland, northern England, Rhône-Alpes, Baden-Württemberg, and Emilia-Romagna. The four nodes previously described display the greatest rates of increasing specialisation.

Regional specialisation, nevertheless, does not affect a single area uniformly. Other studies have revealed a spatial 'sorting' or localisation of production at a sub-regional or local level. Within the region of Poitou-Charentes, in western France, for example, cereals have become more localised in the east in the *département* of Vienne, and vines in the south of Charente around Cognac.[28] Moreover, even within the *département* of Deux-Sèvres, wheat and barley increasingly characterise farming on the calcareous soils of the plain of Thouars,[29] whereas livestock farming (beef and milk) has increased in importance in the north and west. Another typical result of the process of specialisation at the local level can be found in the Calvados *département* in the Basse-Normandie region of northern France.[30] Prior to the 1960s agriculture here had four regional divisions: Le Bocage Virois – a region of small farms (average 21 hectares) with mixed-livestock rearing; La Plaine de Caen–Falaise – a lowland area with large farms (average 38 hectares) and fields, producing cereals together with sugar beet and flax; Bessin and Pays d'Auge – two separate regions characterised by high-quality grassland, a variable farm-size structure (average 27 hectares) and milk (butter and cheese) with some beef production. Between 1955 and 1970, the number of farm units fell by twenty per cent in the region, especially farms under ten hectares in size. At the same time, agriculture throughout the region became specialised in milk production under the stimuli of guaranteed prices, producer co-operatives, and the diffusion of modern dairy technology. On the small farms of the region, dairying became the only economically viable enterprise. On the Plaine de Caen, however, the large farms became even more specialised in the production of cereals, especially wheat, while the smaller farms maintained their profitability by expanding the production of beef and using their cereals for livestock feed rather than selling the grain as a cash crop. Thus not only did the region become more specialised in dairy production, but within the region the distinction between the cereal/beef plain and the surrounding dairy regions was drawn more distinctly.

Conclusion
This chapter has shown how the process of regional specialisation is occurring

for both crops and livestock products, and irrespective of the type of market support provided by the CAP. In addition, the results of the process have been identified at national, regional, and sub-regional levels of analysis. At the national level, three groupings of countries can now be distinguished which have similar trends in the changing national specialisation of production (Table 35). There are, however, some variations in detail within each group. France and West Germany form Group I and have developed a specialisation in products such as barley and sugar beet at the expense of potatoes and milk. The three countries which joined the Community in 1972 form group II. A rising specialisation in wheat, barley, milk and poultrymeat has been compensated by a decline in the relative importance of products such as beef and pigmeat. The remaining four Member States comprise Group III, with a rising specialisation in beef, milk, and pigmeat, and a falling significance for wheat, eggs and poultrymeat.

Table 35. Changing national specialisation of agricultural production, 1963–81[a]

	Group I	Group II	Group III
Countries	West Germany	Ireland	Belgium
	France	United Kingdom	Italy
		Denmark	Luxembourg
			Netherlands
Trends (Increasing Specialisation[b])			
	barley	wheat	barley
	maize grain	barley	sugar beet
	sugar beet	poultrymeat	beef
	poultrymeat	milk	pigmeat
			milk
Trends (Decreasing Specialisation[b])			
	potatoes	potatoes	wheat
	vegetables	sugar beet	potatoes
	pigmeat	vegetables	eggs
	milk	beef	vegetables
	eggs	pigmeat	fruit
		eggs	poultrymeat

a: United Kingdom and Denmark 1972–81, Ireland 1975–81; b: Proportions of the total value of national output over fourteen types of crops and livestock
Source: author's calculations from agricultural statistics

It has been established in the last two chapters that changes in national agricultural production result from even more marked regional trends towards specialisation. Such trends appear to be based on regional cost advantages existing prior to the introduction of the CAP. They are caused by variations in the physical environment, relative location with regard to urban markets and ports for imported cereals, farm-size structure, and the efficiency of marketing and food processing in each region. Unfortunately, the nature of regional cost advantages has to be inferred since data on comparative regional production costs for each farm product are not published. There is a clear need for comparative regional studies on this topic as there is for studying production-cost differences between the Member States themselves. Nevertheless, the creation of a large market in agricultural products, together with a common organisation of the market for each product, has accentuated the regional pattern of cost advantages, but no simple correlations exist between CAP measures and agricultural trends.[31]

Underpinning these regional trends in agriculture have been fundamental changes in the structure of production at the farm level. Individual producers have been giving up the production of their minor or subsidiary enterprises, especially in regions where they possess no particular advantage in production costs. By concentrating capital, land and labour resources on a limited range of products, economies of scale can be obtained. At the same time smaller producers have been leaving agriculture altogether so as to seek more profitable employment in urban areas. Their land has been purchased by owners of larger farms so that the production of all commodities has become concentrated on fewer but larger holdings. The degree of concentration varies from product to product but is most severe in the intensive livestock sector (Table 36). Of holdings with pigs, for example, 8 per cent contain 75 per cent of all pigs in the Community, while fewer than 1 per cent of holdings with poultry account for 73 per cent of broiler (poultrymeat) production. In Brittany, to take a regional example, 51 per cent of producers gave up producing pigs between 1971 and 1980, although production increased by 65 per cent, mainly in herds with more than two hundred pigs.[32] To some observers[33] these structural changes at the farm level have had a greater impact on raising agricultural production than factors such as the increasing yields of crops and livestock or improvements in feed conversion ratios. Clearly, the themes of concentration and specialisation can be analysed at national, regional and farm levels, and are indicative of deep-seated agricultural changes that occur in all industrial market economies.

One consequence of these developments for the CAP is the opportunity to revise policy measures so as to take into account the greater regional

Table 36. The concentration of production by enterprise in the EC, 1977

Enterprise	% largest holdings[a]	% ha or livestock numbers	% holding with the enterprise
Pigs	7.6	75	37
Broilers (fowl)	0.5	73	26
Cattle (beef)	19.0	59	50
Laying fowl	0.2	56	50
Sugar beet	14.0	56	6
Potatoes	9.2	54	33
Cereals	8.3	53	61
Vegetables (total)	5.8	49	31
Sheep	4.6	48	10
Dairy cows	12.0	44	35
Fruit (and berries)	7.5	41	10

a: various definitions
Source: Commission of the European Communities, *The Agricultural Situation in the Community, 1982 Report*, (Brussels, 1983), pp. 88 and 325

concentration and specialisation of production. A more explicit regional pricing policy becomes possible when a product is regionally localised rather than dispersed within the Community. Precedents for the regionalisation of policy measures have already been established, for example, in the allocation of production quotas for sugar beet, in the application of regulations, including prices, governing the production of table wine, in the provision of income support under the LFA Directive, in the delimitation of areas eligible for particular grant-aided schemes, such as the measures for the Mediterranean regions, and in the provision of market support for products with limited spatial distributions, such as tobacco, rice and durum wheat. Moreover, schemes for the regionalisation of prices for major products such as milk have already been advanced.[34] The level of market support, and measures aimed at fostering structural changes in production, could be varied according to the particular character of agricultural production in each of a defined set of regions. This would give specific recognition to variations in farm-size structure and the combination of different products in each region. To date, the CAP largely ignores the marked regional contrasts that characterise agriculture in the EC.

Notes

1 Commission of the European Communities, *The milk and beef markets in the European Community*, Information on Agriculture 10 (Brussels, 1976), p. 36
2 Commission of the European Communities, *The milk and beef markets*, p. 211.

3 M. Butterwick and E. Neville-Rolfe, *Food, farming and the Common Market*, (London, Oxford University Press, 1968), pp. 115–16.

4 Agriculture Economic Development Committee, *Milk production in the European Community: a comparative assessment* (London, National Economic Development Office, 1981), p. 22.

5 A. H. Kampp, *An agricultural geography of Denmark*, (Budapest, Akademiai Kiado, 1975), p. 55.

6 Small farms are exempt from the co-responsibility levy but a 'super' levy was introduced for all producers in 1984.

7 Centre for Agricultural Strategy, *The efficiency of British Agriculture*, Report 7, CAS (University of Reading, 1980), pp. 53–6; Agriculture E.D.C., *Milk production in the EC*, p. 10.

8 Commission of the European Communities, *The milk and beef markets*, p. 39.

9 Centre for European Agricultural Studies, *The EEC milk market and milk policy* (Wye College, University of London, 1977), p. 11.

10 Food and Drink Industries Council, *The food industry and the farmer*, Occasional Paper, FDIC (London, 1982), p. 13.

11 J. Boichard, 'Le lait et les problèmes de l'élevage laitier en France', *Revue de Géographie de Lyon* XLVII (1972), pp. 99–135.

12 L. Hubbard, *Herd size and impact of reducing EEC dairy support prices*, Discussion Paper 5 (Department of Agricultural Economics, University of Newcastle-upon-Tyne, 1982), p. 15.

13 A. C. Byrne, 'The beef and veal situation in France', *Quarterly Review of Agricultural Economics* XXVIII (1975), pp. 80–86.

14 The premium is paid under Regulations 1357/80 and 1417/81. In 1982 its value in the United Kingdom was £12.37 per cow (£24.74 in Northern Ireland). Breeding herds in all locations are eligible, but dairy cows must be crossed with a beef bull and no milk deliveries can be made from the farm.

15 I. R. Bowler, 'Regional agricultural policies: experience in the UK', *Economic Geography* LII (1976), pp. 267–80.

16 Byrne, *Quarterly Review of Agricultural Economics* XXVIII (1975), pp. 80–6.

17 J.-P. Direy, 'L'industrie française de l'alimentation du bétail', *Annales de Géographie* LXXXVIII (1979), pp. 671–704.

18 R. Behrens and H. Haen, 'Aggregate factor input and productivity in agriculture: a comparison for the EC member countries 1963–76', *European Review of Agricultural Economics* VII (1980), pp. 109–146.

19 Kampp, *An agricultural geography of Denmark*, p. 55.

20 These divisions are reflected in the separate payments of the Hill Livestock Compensatory Allowance, under Directive 75/268, between pure-bred and cross-bred ewes.

21 J. W. Alexander, *Economic Geography* (New Jersey, Prentice-Hall, 1963), p. 131.

22 S. Van Valkenburg, 'Land use within the European Common Market', *Economic*

Geography XXXV (1959), pp. 1–24; and 'An evaluation of the standard of land use in Western Europe', *Economic Geography* XXXVI (1960), pp. 283–95.

23 W. C. Found, *A theoretical approach to rural land-use patterns* (London, Edward Arnold, 1971), pp. 57–82.

24 R. Belding, 'A test of the von Thünen locational model of agricultural land use with accountancy data from the EEC', *Transactions of the Institute of British Geographers* VI (1981).

25 G. N. Minshull, *The New Europe* (London, Hodder and Stoughton, 1980), pp. 96–8.

26 R. Lösch *et al*, *Die landwirtschaft in den regionen der EWG und ihre verbindung zu den anderen wirtschaftsbereichen*, Studien zur agrarwirtschaft 8 (Munich, Info-Institut fur Wirtschaftsforschung, 1971).

27 P. Rainelli and F. Bonnieux, *Situation et evolution structurelle et socio-economique des régions agricoles de la Communauté*, Information sur l'agriculture 52 (Brussels, Commission of the European Communities, 1978).

28 J. Pitie, 'L'évolution agricole en Poitou-Charentes dans le siècle' *Norois* CXIII (1982), pp. 141–54.

29 B. Vergneau, 'L'élevage bovin et les transformations récentes des systèmes de productions agricoles en Deux-Sèvres', *Norois* CII (1979), pp. 161–180; G. Bernard, 'Les transformations actuelles de l'agriculture de la plaine du Thouarsais', *Norois* CIX (1981), pp. 45–54.

30 P. Alphandery, *Trente ans d'unité: le syndicalisime gestionnaire dans le Calvados*, Rapport 3, Institut National de la Recherche Agronomique, (Rennes, 1977).

31 P. Henry, *Study of the regional impact of the common agricultural policy*, Regional Policy Series 21 (Brussels, Commission of the European Communities, 1981).

32 C. Canevet, 'Le recensement général de l'agriculture de 1980', *Norois* CXI (1981), pp. 413–19.

33 A. Maris and J. De Veer, 'Dutch agriculture in the period 1950–70 and a look ahead', *European Review of Agricultural Economics* I (1973), pp. 63–78.

34 Centre for European Agricultural Studies, *The EEC milk market and milk policy*, (Wye College, University of London, 1977), p. 38. Four regions can be delimited: 1: specialist dairy, 2: unfavourably structured with alternatives to dairying, 3: cattle the only alternative to dairying, 4: well-structured with alternatives to dairying.

9
Agricultural trade under the CAP

Previous chapters have shown how agriculture has become more concentrated and specialised in Member States and regions of the Community. The CAP appears to have exerted an indirect rather than direct influence, although the creation of a common market has played a more positive, enabling role in these developments: an increased regional specialisation of production has been encouraged and facilitated by the removal of internal trade barriers resulting in the increased competitive trading of agricultural products within the EC. In these circumstances, the rising volume of intra-EC agricultural trade becomes a useful additional indicator of the locational changes taking place in agriculture under the CAP, and shows 'revealed comparative advantage'.[1] Unfortunately, regional trade statistics are not published, and the analysis has to be confined to the countries of the Community. The EC also operates a variable levy system for a wide range of products. The system protects the domestic market from fluctuations on world markets and maintains higher prices within the Community. The concepts of 'stability' and 'security of supplies' have been used to justify such protection in the context of international trade. Consequently, agricultural developments inside the EC, together with the level of protection and concessions that might be made on imports, have an impact on non-member countries. All of these aspects are now examined.

Intra-EC trade in agricultural products
A distinction has already been drawn between trade creation and trade diversion when a common market is established (see Chapter 1), while a division also exists between short-run adjustments in trade patterns and long-run dynamic effects brought about by the permanent alteration of economic variables under the process of economic integration. The

quantitative measurement of each type of effect is fraught with difficulty, but the removal of internal barriers to agricultural trade, together with the import levy system, has undoubtedly resulted in substantial trade diversion for temperate-zone agricultural products, especially after 1965,[2] with some evidence of overall trade creation.[3] The value of agricultural trade amongst the countries of The Nine rose from 2,208 million EUA in 1958 to 36,343 million EUA in 1980, a marked increase even allowing for the effects of price inflation.[4] The EC share of all imports of agricultural products also rose from 23 per cent in 1959, through 39 per cent in 1969, to over 45 per cent by 1979. Only non-EC imports of animal feed, fruit and vegetables continued to increase appreciably in the early years of the Community. While the share of non-EC countries in the import trade of the Community fell, the value and volume of such trade nevertheless rose substantially – from 7,190 million to 42,210 million EUA between 1958 and 1980. These increases in agricultural trade must be set against the background of a declining relative importance for agriculture in the total trade of the Community. In 1958, for example, agricultural commodities accounted for 26 per cent of the value of all trade by The Six (imports and exports); by 1973 the proportion had fallen to 18 per cent for the same group of countries, and to 12 per cent by 1981 for The Nine.

West Germany, Italy, the United Kingdom and France are the principal agricultural importers amongst the Member States, in aggregate accounting for over three-quarters of imports from all sources. Dependence on the EC for food imports, however, is variable and ranges from 74 per cent for Ireland (in 1981) to 39 per cent for Denmark (EC average 46 per cent). The extent to which countries have re-orientated their food imports towards the other Member States takes on yet a different pattern: in the 1960s, Italy and the Netherlands increased their imports from within the EC to the greatest extent, while in the 1970s Denmark, Ireland and France were the countries most affected. In the case of Italy, the growth in consumer income in the 1960s led to a demand for livestock products which the national agricultural sector was unable to meet due to handicaps of soil, climate, low technology and small farm structure. Interestingly, the United Kingdom has reduced her traditional dependence on food imports and, since joining the Community, has replaced non-EC with EC sources. Nevertheless, the continued volume of imports from outside the Member States (sheepmeat, wheat, sugar) renders the country a considerable contributor of import levies to the Community Budget.

In Table 37, the import substitution effects of the common market are clearly demonstrated for the period prior to the first enlargement of the

Community. The increase in internal trade exceeded that of external trade in all countries for food imports, for products both subject and not subject to import levies (regulated) alike. The incidence of import regulations appears not to have influenced the rate of increase in import trade with non-EC countries, but internal trade grew faster for regulated as compared with non-regulated products. For the EC of The Six, therefore, internal trade in all food imports grew by 278 per cent between 1963 and 1972, whereas imports from other countries grew by only 48 per cent (Table 37).

Table 37. Agricultural trade in The Six, 1963–72 (1972 as a percentage of 1963 by value: m./UA)

Country	Imports				Exports			
	All food		Regulated Products[a]		All food		Regulated Products[a]	
	Int.	Ext.	Int.	Ext.	Int.	Ext.	Int.	Ext.
West Germany	316	153	314	155	630	297	1109	333
France	353	133	374	128	451	189	554	196
Italy	545	152	731	158	263	173	265	195
Netherlands	472	174	765	161	336	161	332	153
Belgium/								
Luxembourg	372	135	526	143	349	186	509	222
EC(6)	378	148	434	150	381	191	439	199

Int.: internal trade between Member States; Ext.: external trade with third countries
a: Subject to import levies and export refunds
Source: author's calculations from trade statistics in Eurostat, *Yearbook of Agricultural Statistics* (various years)

Turning now to the agricultural export trade of Member States, over 65 per cent of exports are destined for other countries in the Community, a proportion that has been rising steadily over the last two decades. Belgium/Luxembourg, Ireland and the Netherlands are the major exporters, all sending nearly three-quarters of their agricultural exports to other Member States; proportions have risen most rapidly in Italy, West Germany and the United Kingdom in recent years.[5] As with food imports, the volume of exports increased at a faster rate amongst the Member States (281 per cent 1963–72) than with non-member countries (91 per cent), and was greater for regulated as compared with non-regulated products (Table 37). Thus the principle of 'Community preference' had a clear impact on the growth of agricultural trade and market integration within the EC but in a way that varied between the Member States. There appears to have been a selective reorientation of traditional patterns of internal trade as imports from outside

Table 38. Intra-EC trade in agricultural products, 1967–80 (by weight)^a

Product	Volume (1980/1)^b	Percentage change 1967/8–1971/2 (The Six)	1974–1980 (The Nine)
Wine	21,792^c	+36.5	+7.3^d
Common wheat	5,850	−29.1	−1.3^d
Barley	4,465	+4.1	+0.2^d
Fresh vegetables	4,024	+6.9	+5.3^d
Maize	3,878	+58.2	+0.4^d
Fresh fruit	3,757	+6.1	n.d.
Soya cake	2,356	+29.8	+10.0
Potatoes	2,436	+13.2	+7.0
Pigmeat	1,994	+22.4	0
Beef and veal	1,297	+3.1	+3.1

a: ten most important products by weight traded; b: '000 tonnes; c: '000 hectolitres; d: 1973–9

Source: Commission of the European Communities, *The Agricultural Situation in the Community, 1983 Report*, (Brussels, 1984), pp. 241–2

the EC have been displaced by Community products.[6]

Only a limited number of products are traded in significant volumes within the Community (Table 38). Wine and common wheat are the main commodities, followed by barley and fresh vegetables. Trade in beef (and veal), the tenth-ranked commodity by weight traded, amounts to only 1.3 million tonnes each year, a relatively small proportion of all the beef produced within the EC. However, the rate of increase in the volume of produce traded shows two interesting trends. First, livestock feeds (maize, soya cake, common wheat) exhibited high rates of increase in the 1960s reflecting the developments in the intensive-livestock sector that have been discussed in previous chapters. Increasing levels of trade have been maintained in the 1970s only for soya cake; the high rates of increase in intra-EC trade in sugar, cereals and oils/fats, so characteristic of the early years of the Community, have not been continued. Secondly, the dynamism of all agricultural trade within the EC appears to have been lost since the mid-1970s. Compared with earlier years, rates of increase in trade have been modest and for some commodities the volumes traded have been static or even in decline (Table 38).

One reason for the lower rates of increase in EC-trade in recent years is contained in Table 23. As national levels of self-sufficiency have risen for an increasing number of products, so the scope for intra-EC trade has diminished. Milk and dairy products, sugar and cereals have been particularly affected. In addition, Table 23 gives some indirect guidance on the countries which act as major suppliers or consumers of specific agricultural products in intra-EC trade. The Netherlands, for example, exhibits declining levels of self-sufficiency in barley and fresh fruit, but has a growing surplus of dairy products, sugar, eggs and pigmeat for export. West Germany, France and Italy, by contrast, have experienced falling levels of self-sufficiency in pigmeat and fresh vegetables which lead toward a higher level of imports.

The direct evidence on trade flows from international trade statistics reveals a complex web of imports and exports on a commodity-by-commodity basis. To simplify the situation, aggregate exchanges of agricultural produce are examined in Table 39. In part A, the rows of data show the destination of exports from each country to the other Member States. Most countries have three main trading patterns: France, for example, exports mainly to Italy, West Germany and Belgium/Luxembourg. There are, however, two anomalies to the pattern. On the one hand Ireland and Italy have one major trading partner each: the United Kingdom in the case of Ireland, and West Germany for Italy. On the other hand the United Kingdom has a diversified pattern of export links, trading more equally with five other countries in the Community.

The pattern of agricultural imports (Table 39B) shows a strong degree of reciprocity with exports. Belgium, for instance, imports agricultural products mainly from the Netherlands and France, while the same two countries receive a majority of Belgian exports. Possibly the strongest inter-dependencies are between Ireland and the United Kingdom, and between Denmark and West Germany. There are, nevertheless, deviations from the general pattern of reciprocity, the most noticeable being the cases of Italy and the United Kingdom. While Italy exports strongly to West Germany, most of the country's agricultural imports are drawn from France. Similarly, a considerable proportion of the United Kingdom's exports go to West Germany, but imports are drawn more from Denmark (pigmeat) and the Netherlands (dairy products). The degree of dependence on a few trading partners is similar between the patterns of imports and exports in intra-EC trade. Most countries obtain their imports from two or three Member States, although Ireland is again very dependent on imports from the United Kingdom; the latter country, together with France, has a diversified pattern of trading links for food imports.

In aggregate, therefore, intra-EC trade in food products has a relatively simple structure even though individual commodity flows are more complex

Table 39A. Exports of agricultural produce to other Member States, 1976–7 (% all exports by value)[a]

Country of origin	B/L	D	F	WG	Ir	It	N	UK		Index of trade concentration[b]
				Country of destination						
Belgium/										
Luxembourg (B/L)	–	0.5	31	25	0.3	7	31	6	(100)	.263
Denmark (D)	3	–	8	31	0.3	16	5	37	(100)	.268
France (F)	20	1	–	25	1	29	10	15	(100)	.219
West Germany (WG)	11	6	21	–	0.3	31	23	9	(100)	.217
Ireland (Ir)	6	0.4	12	7	–	2	10	62	(100)	.418
Italy (It)	5	2	29	49	0.4	–	4	10	(100)	.339
Netherlands (N)	17	1	16	46	0.6	9	–	11	(100)	.286
United Kingdom (UK)	12	3	23	21	17	9	15	–	(100)	.172

B. Imports of agricultural produce from other Member States, 1976–7 (% all imports by value)[a]

Country of destination	B/L	D	F	WG	Ir	It	N	UK		Index of trade concentration[b]
				Country of origin						
Belgium/										
Luxembourg	–	2	36	12	3	4	37	7	(100)	.289
Denmark	3	–	13	48	1	7	15	13	(100)	.293
France	22	4	–	18	4	15	27	11	(100)	.192
West Germany	9	8	20	–	1	14	43	5	(100)	.262
Ireland	2	1	15	2	–	2	9	68	(100)	.494
Italy	5	8	40	27	0.8	–	16	4	(100)	.269
Netherlands	31	3	20	29	5	3	–	10	(100)	.235
United Kingdom	4	19	22	9	21	6	19	–	(100)	.178

a: there are rounding errors in the data;

b: $\Sigma \left(\dfrac{xi}{\Sigma xi} \right)^2$ where xi is the value of exports/imports with each country

(0.14 = diversified; 1.0 = concentrated)

Source: Economic Commission for Europe, *Agricultural trade in Europe*, Agricultural Trade Review 16 (New York, United Nations, 1979), p. 16

and are not considered in detail here. Nevertheless, the trade pattern of West Germany requires particular attention, for it exhibits once again the influence that can be exerted by national inflation and currency exchange rates. West German agricultural exports have risen significantly in the 1970s, especially for milk, sugar and butter; West Germany's share of intra-EC sugar exports, for example, rose from 5 to 26 per cent between 1971 and 1977, while comparable figures for butter were 9 and 26 per cent. The rise of West German agricultural exports can be attributed to many causes including a good marketing organisation, an efficient advisory service, and a flexible, regionalised banking system. Nevertheless, an over-riding cause has been the 'distortions' in competitive conditions introduced through the agri-monetary system and MCAs.[7] Three effects can be identified: higher farm prices in West Germany than elsewhere have encouraged farm production; exports from West Germany have attracted positive MCAs which act as export subsidies; purchased inputs are made in real money rather than 'green' money, and the cost of these inputs, relative to the prices which the farmers receive for their products, have been lower in West Germany. Thus, West German agricultural exports attracting MCAs have increased at a greater rate than those without MCAs, while the converse has occurred for imports.[8] MCAs, therefore, have tended to distort patterns of agricultural trade in the 1970s in favour of those countries with strong currencies.

Extra-EC trade and agricultural products
Most expert opinion concludes that there has been more trade diversion than trade creation within the EC,[9] although the situation has been complicated by dynamic changes in the total volume of agricultural trade over the last two decades. In addition, the Community has negotiated a highly complex set of commercial agreements with a wide range of countries which has been termed a 'pyramid of privilege'.[10] Thus, before examining the Community's pattern of external trade, it would be helpful to set out the agreements that have been made with non-member countries, recognising that they reflect particular foreign policy objectives as much as commercial considerations.[11]

The pyramid of privilege:
At the base of the 'pyramid' is a Generalised System of Preferences (GSP) which gives over 380 processed agricultural products entry to the EC market either free of customs duties up to certain quantitive limits, or subject to low rates of duty. The Common External Tariff (CET) can be applied to 'excess' imports, while the GSP is non-reciprocal in nature. The System has been applied by the EC since 1971 to all developing countries and arose out of negotiations in the United Nations Conference on Trade and Development (UNCTAD). A number of 'sensitive' products, however, are excluded from preferential treatment, while the limits are often judged to be too restrictive for real assistance in the economic development of Third World countries.[12]

In addition, the EC has taken part in negotiations under GATT since 1963, including the Kennedy (1963–7) and Tokyo rounds (1973–9). The Community accedes to the aim of ensuring that subsidies to agriculture should not lead to a country having more than an equitable share of the world market. The exact definition of 'equitable' remains in dispute with other exporters on the world market, while the EC in any event has made only modest concessions in agricultural trade under GATT, and then mainly for tropical products, animal feed (soya, cassava, maize gluten) and oilseeds. Nevertheless, the EC claims that the Tokyo round implicitly recognised the principles and mechanisms of the CAP, a conclusion that is also contested, particularly by agricultural exporting countries such as the United States. The EC has been party to a number of other multilateral agreements which have had the objective of stabilising world markets; these include the International Wheat (1971), Coffee (1976), and Cocoa (1970) Agreements. Member States are often represented in their own right in these negotiations.

A set of bilateral, non-preferential trade or Commercial Co-operation agreements (CCAs) have been signed with a more limited number of countries. Their objective has been to promote the development and diversification of trade, including some specific products such as beef. The countries involved include India (1973), Mexico (1975), Pakistan and Bangladesh (1976), China (1978) and Yugoslavia (1980). Some of the agreements were designed to meet the problems created by the loss of Commonwealth Preferences when the United Kingdom joined the EC.

Moving further up the 'pyramid', bilateral, preferential trade agreements have been signed with many developing countries including some former overseas territories of the Member States. These agreements allow the trading partners access to the EC market for agricultural produce under concessionary terms, although usually for processed products not covered by the CAP. Mediterranean countries form the major element in these agreements and have included Spain (1970), Portugal (1972), Algeria (1976), Tunisia (1976), Morocco (1976), Egypt (1977), Jordan (1977) and Syria (1977). Since the exports of these countries compete directly with the produce of some of the poorest regions of the EC, there are often seasonal restrictions on the imports of 'sensitive' products such as tomatoes, cucumbers, melons and onions. In addition, agreements exist with the seven countries comprising EFTA which are important partners in terms of the volume of trade.

Yet more favourable trading privileges have been accorded to a large group of developing countries in Africa, the Caribbean and the Pacific (ACP

States). They are mainly former colonial territories of the Member States and have risen in number from the first Conventions signed at Yaoundé in 1963 and 1969, and later at Lomé in (I) 1975, and (II) 1979. There are sixty-four of these states today. The Lomé Conventions are seen as the cornerstones of the EC's relations with the Third World, the ACP states having duty-free access to the Community market for almost all their tropical agricultural products with non reciprocity for the Member States. The EC has only most-favoured-nation status. Sixty per cent of the agricultural exports of the ACP countries falling under the CAP are also effectively granted free access to the EC,[13] while an annual market for 1.3 million tonnes of sugar has been guaranteed since 1975. In addition, a Stabex (1976) scheme to stabilise the export earnings of the ACP countries has been applied to forty-four primary agricultural products. These include cocoa, coffee, groundnuts, tea and sisal, but exclude products subject to the CAP. If, during a calendar year, an ACP state's exports of an eligible commodity falls below a 'reference' level (usually the state's average sales to the EC in the preceding four years) by a predetermined margin,[14] that country can request a transfer from the Community's stabilisation fund to cover all or part of the gap in export earnings. Transfers mainly take the form of interest-free loans repayable over five years, but the poorest countries are usually relieved of having to repay the loan. In Table 40 a range of countries, with their dependency on the export of a few primary products, are listed. Burundi (coffee), Uganda (coffee), Chad (cotton), Ghana (cocoa) and Tonga (copra), for example, are highly dependent on the EC for export earnings. In the late 1970s, the benefits of the Stabex scheme became very concentrated on a few products such as wood, coffee and groundnuts, and on a few countries including Senegal, Ivory Coast and Sudan. In addition, the funds set aside for the scheme proved inadequate,[15] especially in 1981, while Stabex now competes with a similar compensatory facility provided by the International Monetary Fund.

At the apex of the pyramid are a few countries that have entered into bilateral Association Agreements with the EC. These arrangements were intended to lead ultimately to full membership of the Community, including the gradual removal of duties in trade and the step-by-step alignment of tariffs to the CET. Greece (1961) made the step to full-member status in 1981, but Turkey (1963), Malta (1970) and Cyprus (1973) have yet to complete the transition and now seem unlikely to do so.

Thus a mosaic of more than twenty-five agreements has been built by the Community with some ninety countries throughout the world in the

Table 40. ACP states and the Stabex system 1975

Countries	Proportion of Stabex products in total exports	Global percentage of exports covered by Stabex
Five most dependent on exports of Stabex products:		
Burundi	coffee (86%), cotton (3%), leather (6%)	95
Gambia	groundnuts (94%)	94
Uganda	coffee (66%), cotton (15%), tea (5%)	86
Ghana	cocoa (61%), timber (19%)	80
Guinea Bissau	groundnuts (69%), timber (5%)	77
Five least dependent on exports of Stabex products:		
Sierra Leone	palm kernel oil (5%)	15
Botswana	leather and hides (9%)	9
Fiji	coconut oil (5%)	5
Jamaica	bananas (4%)	4
Swaziland	cotton (3%)	3

Source: C. Cosgrove Twitchett, *Europe and Africa: from association to partnership*, (Farnborough, Saxon House, 1978), p. 153

matter of agricultural trade. The agreements appear haphazard in their development reflecting a step-by-step approach by the Community to external pressures rather than the pursuit of a clear goal. Nor is the above list of agreements exhaustive: it does not contain the special arrangements made by the EC for the importation of specified quantities of New Zealand butter, for example, nor the bilateral agreements with the United States on imports of high-quality beef, turkey, table-grapes and prunes in return for concessions on EC exports of cheese, spirits and beef. In so far as a variety of concessions has been made, however, so a pyramid of privilege exists in the external trade relations of the EC. Countries at the 'top' of the pyramid often see concessions, such as those of the GSP, as eroding their trading advantages in the EC market.

The pattern of trade

The EC is the world's largest food importer (26 per cent of all imports), and between 1973 and 1980 the value of imports from non-member countries rose by 97 per cent. The volume and pattern of agricultural imports still reflect past developments in Western European agriculture. About half consist of tropical or semi-tropical items which cannot be produced in the Community or to any great extent (rice, tropical fruits, tobacco, coffee), while the importation of 'temperate' products (barley, butter, beef, sheepmeat) has declined with rising levels of self-sufficiency in the Community. Nevertheless, the Member States remain important world markets for oilseeds (soya), wine, feed grains (maize) and meat (pigmeat) (Table 41). Coffee, cocoa, tea and spices comprise the most important group of imports from the ACP states, while fruit and vegetables are the principal imports from the Mediterranean countries. It should be noted, however, that relatively small proportions of world production are traded in most 'temperate' products ('residual quantities'), and only soya, sugar and SMP are significant in this respect.

Table 41. The role of the EC in world agricultural trade of 'temperate' products, 1980 (% world trade in each product)

Product	Imports	Exports	% world pro- duction traded
Total cereals	8.3	8.8	15.9
: wheat	5.2	13.4	20.1
Feed grains	11.1	4.7	13.4
: maize	13.8	0.2	17.5
Oilseeds	43.0	0.1	16.6
: soya	43.2	0.0	30.3
Wine	20.7	35.7	7.9
Sugar	5.9	13.8	27.0
Milk	0.5	60.1	0.0
Butter	13.1	57.4	12.3
Cheese	12.7	44.4	6.3
SMP	0.1	60.0	29.9
Total meat	12.9	17.0	3.8
: beef & veal	6.4	17.6	5.2
: pigmeat	9.7	20.1	1.1
: poultrymeat	4.1	29.2	4.4
Eggs	2.5	19.6	1.4

Source: Commission of the European Communities, *The Agricultural Situation in the Community, 1983 Report* (Brussels, 1984), p. 260

France, West Germany, Italy, the Netherlands and the United Kingdom are the main importers of food from outside the Community, although in recent years rates of increase for such imports have been highest in the Netherlands, Belgium/Luxembourg and Denmark. The greatest imports by volume and value come from other developed countries, especially the United States (Table 42). This feature reflects an important structural element of international trade, namely, that a major proportion of the exchange of food products is between developed countries. Nevertheless, imports from developing countries (45 per cent of agricultural imports), especially the ACP states and Latin America, have been increasing at a faster rate than intra-EC imports in recent years, principally at the expense of imports from Mediterranean, other West European and State Trading countries. These developments continue trends established in the 1960s and can be offered as encouraging, if not strong, evidence of positive effects created by the trade agreeements between the Community and developing countries.

Table 42. Trading partners of the EC in agricultural produce, 1973–81^a (% total trade by value with each partner)

Country	Imports		Exports	
	1973	1981	1973	1981
United States	17.3	20.7	16.6	10.1
Latin America	15.7	16.3	3.5	4.4
Industrial Commonwealth	14.1	12.9	4.6	3.3
Mediterranean	11.6	9.7	14.5	17.4
ACP States	10.9	11.7	7.8	10.1
West Europe	10.3	7.4	23.4	17.8
State Trading	10.3	7.4	10.1	12.1
Rest	9.8	13.9	19.5	24.8

a: The Ten
Source: Commission of the European Communities, *The Agricultural Situation in the Community, 1982 Report*, (Brussels, 1983), p. 253 and 248–9

Besides being an important world market for food products, the EC is increasingly important as an exporter (10 per cent of world exports). Between 1973 and 1980, for example, the value of exports rose by 189 per cent – faster than the growth of food imports. France, West Germany, the Netherlands and the United Kingdom are the main exporting countries of the Community, although Ireland and Italy have experienced the greatest increases in agricultural exports in recent years. Two-thirds by value of exports are in the form of processed food and drinks and are less focussed on other developed countries than agricultural imports (Table 42). Dairy products (butter, cheese, SMP), flour, sugar and canned meats, for example, are exported to developing countries; high-value products such as alcoholic drinks (wine, whisky), cereal products (biscuits, confectionery) and fruit/vegetables are sent to other industrialised countries; whereas alcoholic drinks, refined sugar and condensed milk feature in trade with less-developed European countries such as Yugoslavia and other Mediterranean states. The export trade with the Communist bloc is conducted mainly between Greece, West Germany and France, on the one hand, and the Soviet Union and Hungary on the other, and includes products such as fruit, vegetables, oils/fats, cereals and sugar.

After cereals (common wheat, barley), sugar is the single most important item by volume exported by the EC, although the main increases in exports in recent years have been in butter/butteroil, whole milk powder, sugar and wine. Such exports account for considerable proportions of world trade in the products (Table 41). Only pigmeat of the major commodities shows a decline in export volume in recent years, and again trade with the United States and Western Europe has declined relative to exports to Mediterranean and ACP states (Table 42).

The budgetary cost

A considerable proportion of EC exports attracts a price subsidy since in most years world market prices tend to lie below those within the Community (Table 43). Moreover, as foreign observers frequently comment,[16] the decision to export often appears to be based primarily on the existence of Community surpluses, with seemingly little regard for the supply and price situation on world markets and usually despite the lack of any competitive advantages due to lower production costs. As has been explained earlier (Chapter 3), producers of milk, cereals and sugar in northern parts of the Community have been politically successful in establishing and then maintaining very favourable CAP price levels in relation to world prices over the last two decades. By comparison, producers of Mediterranean products have been less politically adroit at exploiting the decision-making structure of the Community in relation to price levels for their products.

Consequently, dairy products, cereals (hard and soft wheat) and white sugar absorb a high proportion of the funds available for export subsidies

Table 43. CAP prices as a percentage of world prices, 1968/9–1980/1[a]

Product	1968/9	1971/2	1974/5	1977/8	1980/1
Skim milk powder (SMP)	365	112	139	494	204
Butter	504	171	316	388	286
White sugar	355	186	41	255	85
Barley	197	185	107	206	134
Wheat (hard)	214	254	120	218	138
Maize	178	176	106	203	147
Beef and veal	169	133	162	196	190
Wheat (soft)	195	209	107	216	146
Oilseeds	203	147	80	153	168
Pigmeat	134	131	109	137	135
Eggs	137	162	164	–	–

a: entry price to EC as % annual average third country offer price
Source: Statistical Office of the European Communities, *Yearbook of Agricultural Statistics* (Brussels, various years 1974–80)

(restitutions) under the Agriculture Fund. The proportion of the Guarantee Section of FEOGA spent on export refunds has been rising steadily in recent years, and stood at 35 per cent in 1982 (Table 44). The greatest expenditure is made subsidising the export of dairy products, followed by cereals and sugar. While cereals and dairy products were also the dominant products in the pattern of expenditure in earlier years, support of the internal market, rather than export subsidy, was the predominant use of the Fund. The recent increased importance of export restitutions, and the polarisation of expenditure on a few products, reflects the nature of the growing problem of agricultural surpluses in the Community. Interestingly, an emphasis on export restitutions also existed in the 1960s at which time France, as the Community's major agricultural exporter, gained most from the Agriculture Fund. With other countries now emerging as net exporters, the benefits of the export restitutions are spread slightly more evenly, to the benefit of West Germany in particular.

Table 44. Expenditure under the Guarantee Section of FEOGA (% total each year)

Commodity	1974 ER	1974 I/S	1978 ER	1978 I/S	1982 ER	1982 I/S
Dairy products	11.0	28.3	18.0	28.2	12.3	14.6
Cereals and rice	2.5	10.5	9.8	3.2	8.9	6.2
Sugar	0.3	3.3	7.4	2.7	6.0	4.0
Oils and fats	b	4.5	b	3.7	b	5.8
Beef and veal	1.8	8.7	1.7	5.7	5.2	4.1
Tobacco	b	5.9	b	2.5	b	4.9
Fruit & vegetables	0.6	1.6	0.6	0.6	0.5	6.9
Wine	b	1.3	b	0.7	0.3	4.3
Pigmeat	1.8	0.4	0.4	0.1	0.8	b
Eggs & poultrymeat	0.5	–	0.4	–	0.8	–
(Sub-total:)	:(18.5)	:(64.5)	:(38.3)	:(47.4)	:(34.8)	:(50.8)
Others	12.6		4.1		11.9c	
Monetary Compensatory Amounts	4.4		10.2		2.5	
Total (m./EUA)	3097.9		8672.7		12405.6d	

ER: Export Refunds; I/S: Intervention and Storage
a: including Accession Compensatory Amounts (ACA); b: less than 0.1 per cent; c: of which olive oil – 4 per cent, sheepmeat – 2 per cent; d: ECUs
Source: Commission of the European Communities, The Agricultural Situation in the Community, 1976 Report (Brussels, 1977), pp. 416–17, 1979 Report, pp. 254–5, 1983 Report, pp. 262–3

Impact on non-Member countries

The agricultural and trade policies of the EC can have two types of impact on non-member countries. First, exports by such countries onto the world market can be displaced by the subsidised exports of the EC; secondly 'Community preference' can displace the exports of non-EC countries on the internal EC market. The extent of the impact varies with a country's exports of temperate agricultural produce and the type of trade agreement, if any, that has been signed with the Community.

To countries competing on world markets, the EC's export subsidies often appear to generate 'unfair competition' and periodically lead to accusations of 'dumping'. Indeed, there was some evidence of this in the late 1970s. Faced by rising and politically embarrassing levels of stocks in intervention stores,[17] the Community exported increasing quantities of agricultural products through the world market. In 1976 for example, the 1,135,000 tonnes of SMP in store represented three-quarters of the annual EC production. Through subsidised exports, this volume was reduced to 180,000 tonnes by 1980 but at a high budgetary cost (Table 44). Stocks of other products have also been kept in check by increasing the volume of subsidised exports. While 310,000 tonnes of sugar were exported in 1974, for example, the volume had risen to 2,695,000 by 1978. Comparable figures for wine were 2,316 million hectolitres and 6,346 million hectolitres. Subsidised exports at these levels are bound to have a damaging effect on the trade of traditional exporters on the world market. Community exports of beef, for example, accounted for 5 per cent of world trade in 1977, but had risen to 21 per cent by 1980 and in so doing displaced Australian and Argentinian exports to Egypt, and Uruguayan exports to Ghana. In general subsidised exports from the EC depress the prices available to traditional exporters on the world market and amplify any price fluctuations.[18] In addition, they have raised the Community's share of world trade in agricultural products, albeit at the modest annual rate of 0.2 per cent over the last decade.

Exporters of 'temperate' products also face protective trade barriers in the EC market. No country is free of agricultural protection, however, while the obscurity of many national systems of assistance tends to throw the EC's more obvious import levies into unjustified relief. The central debate lies in the degree of protection rather than in whether protection should exist or not. Most studies of this issue show that over all products the Community does not maintain exceptionally high levels of agricultural protection compared with other countries. Canada and the United States, for example, offer their dairy producers more protection than the EC, while a similar situation has prevailed in Japan for wheat. However, relative levels do vary through time and from product to product,[19] and estimates for the EC reveal that trade diversion through agricultural protection has been greatest for dairy products, eggs, meat, meat products and cereals.[20]

The example of Canada illustrates the effects that can be created by the Community's protectionist policies (Table 45). The proportion of Canada's exports to the Community accounted for by agricultural products fell from nearly half in the period 1955/9, to only 16 per cent in 1980/1. The most severe contraction occurred when the United Kingdom joined the Community, since the two countries historically have been important trading partners. Other trading links have been developed by Canada to compensate for the loss of markets in Western Europe: the Soviet Union, Japan and the United States have become more important trading partners. Imports of agricultural products from the EC (mainly cheese, skins/hides/furs, chocolate/confectionery, tea/coffee), once on a rising trend, have also declined in recent years (Table 45), with the United States remaining the main source of Canadian food imports. Wheat has always been the main agricultural product exported to the EC and this single commodity still accounts for 40 per cent of all exports by value. The proportion has remained relatively unaltered in recent years with most wheat entering the United Kingdom where it accounts for 62 per cent of all imports from Canada. Nevertheless, alternative markets for wheat have had to be developed in China and the Soviet Union.

Table 45. Canada's agricultural trade with the EC, 1945–82 (by value)

Years (average)[a]	Agricultural products as % total exports to EC	Agricultural products as % total imports from EC	Wheat as % total agricultural exports	Cheese as % total agricultural imports
1945–9	49.1	2.0	n.d.	n.d.
1950–4	41.4	4.6	n.d.	n.d.
1955–9	47.8	6.6	n.d.	n.d.
1960–4	38.5	8.5	n.d.	n.d.
1965–9	32.2	9.1	n.d.	n.d.
1970–4	27.3	7.7	44.8[b]	13.5[b]
1975–9	21.2	7.4	37.3[c]	12.2[c]
1980–1	16.2	6.8	40.1	12.9

a: includes U.K. trade in all years; b: 1972–6; c: 1977–9
Source: Agriculture Canada, Canada's trade in agricultural products 1979, 1980 and 1981 (Ottawa, Agriculture Canada, 1982), pp. 18–19 and 26–7

Similar redirections in trading patterns can be identified for the agricultural sectors of other trading partners of the EC. For the New Zealand dairy and sheep industries[21] the impact of the United Kingdom's membership of the Community has been traumatic. The proportion of all agricultural exports from New Zealand going to the EC fell from 68 to 32 per cent between 1965 and 1977. Even greater reductions are evident when individual products are considered. In 1953, for example, 90 per cent of butter and 93 per cent of cheese exports were sent to the United Kingdom market. By 1972, anticipating the United Kingdom's membership of the EC, less than half the aggregate dairy exports went to that market (40 per cent by volume, 46 per cent by value). New markets and new products, such as SMP and casein, have been developed in Latin America and Asia (India and Japan) to compensate for the decline in trade with Western Europe. However, these markets have not been assured and there has been competition with the subsidised exports of the Community. Moreover, the change in export orientation has not been accomplished without major alterations in the structure of the New Zealand dairy sector. There has been a marked decline in the number of milk producers and an increase in average herd size; Jersey cows have been replaced by higher-yielding Freisian cows; wholemilk tanking has taken over from farm cream collection. Clearly, these developments would have effected dairying in New Zealand eventually, but the scale and the speed of change has been accelerated by the effective closure of a large part of the traditional market in the United Kingdom for the export of dairy products. Similar conclusions can be drawn for the New Zealand sheep sector with the diversification of markets towards Japan and the Middle East. Producers have had to develop different grades of lamb (lighter and leaner carcasses) to meet consumer preferences in the new markets.

The impact of the EC's trading regulations are less clear as far as Third World countries are concerned. Of the agricultural exports of developing countries, 30 per cent are directed towards the Community, nearly 60 per cent entering without duty under preferential agreements. A further 34 per cent are subject to low rates of duty and only 7 per cent of exports are subject to import levies (beef, cereals, rice, sugar). Conversely, only 20 per cent of the Community's exports compete on the world market with produce from developing countries while both the number of countries and products effected are limited.[22] The way in which the Community's agricultural trade policy is interpreted, however, depends on whether a 'classical' or 'critical' mode of analysis is adopted.[23] Observers who adopt a 'critical' analysis tend to emphasise the exploitive nature of the relationship between the ACP states and the EC, describing the trends in trade as 'neo-dependence'.[24] Trade agreements are interpreted as merely a shift in the nature of imperialism,[25] with metropolitan domination a cause of the continued under-development of the ACP states. Overall, the political consensus favours the view that EC policies generally act against the interests of developing countries. Nevertheless it is difficult to discern from the available data any strong impact of agreements such as Lomé I and II on trade flows for good or ill. Generally, the size of concessions has been limited and insufficient to shift trade flows to any appreciable extent. On the other hand circumstances do vary from country to country. About half of the agricultural import and export trade of the Mediterranean countries, for example, is with the Community so that the trading relationship has become a major determing factor for the development of agriculture in these countries.[26]

Food aid
Like agricultural trade, food aid in the EC is inescapably linked with agricultural developments under the CAP. On the one hand, food aid has been made possible by the food surpluses generated in part by price policies under the CAP. On the other, the Community has used food aid as a method of disposing of unwanted agricultural produce so that the food involved has not always been that required by the recipient countries. Of course, food aid has been given for other reasons. It has been used politically by the Commission to foster its 'leadership' ambitions,[27] and by the Member States to enhance the influence, image and prestige of the EC amongst Third World countries. In addition, food aid has been provided by the Community for humanitarian purposes to combat famine, hunger, poverty and misery both in the poorest countries of the world and in emergency situations (earthquakes, floods, refugees from military actions).

As in many other spheres, however, food aid is a mixture of national and Community actions, not least because of its involvement with the separate foreign policy objectives of the Member States. Aid passes to recipient countries: multilaterally from the EC through international agencies; bilaterally from the EC; either multilaterally or bilaterally from each Member State. Recipient countries benefit from subsidised food, aid for development (increased demands for food can be met without price inflation), and budgetary support (food aid can be sold on the domestic market to yield 'counterpart funds' for investment elsewhere in the economy). Nevertheless, food aid must be seen in perspective. Although it has grown incrementally to significant volumes, total international food aid comprises less than 5 per cent of all food exports and has declined in relative importance in the context of commercial food imports to developing countries. Even international cereals aid represents less than 20 per cent of all cereals imports into developing countries.

The role of the EC as a donor of food aid through international agencies began with the Food Aid Convention (FAC) of the 1967 International Wheat Agreement (IWA). Under the FAC, a number of cereal producers agreed to provide minimum quantities of grain, usually wheat, wheat flour and rice, and so share the burden previously carried by the United States as the main provider of international food aid. By 1971 the Community was donating 26 per cent of cereals under the FAC, a proportion that had risen to 30 per cent by 1979 (1,280,000 tonnes per year).[28] A new FAC was signed in 1980 under which the EC pledged 1,650,000 tonnes of wheat per year, over half coming through the EC rather than the Member States. The Community is now the second most important contributor after the United States out of twelve donor countries. About one-third of EC food aid is distributed through international organisations such as the World Food Programme (WFP), United Nations Relief and World Agency for Palestinian Refugees, and the International Committee of the Red Cross. Aid is mainly in the form of SMP and butter oil; agencies are well-suited to reconstituting and distributing dairy products as food aid.

Bilateral food aid from the EC dates from 1968 and has to meet one of three criteria: economic development action, health and nutritional action or emergency relief action. Of these, humanitarian relief has acquired more prominence in recent years but remains a small fraction (10 per cent) of all food aid shipments. Generally, food aid is given on a grant basis from year-to-year once requests for aid have been matched with available resources. The EC covers the cost of the food, the export refunds and, for the poorest countries, transport charges to the border of the recipient country. The objective of bilateral food aid has to be specified by the recipient especially if it is to be sold on the domestic market. This places Community aid on a different basis to the 'project' approach of the multilateral WFP and the 'programme' approach of the United States.

Three principles constrain the disposal of surpluses as food aid: additionality, orderly disposal and voluntary consultation.[29] The first principle encompasses concepts such as 'usual marketing requirements' and 'normal pattern of trade', for food aid should be over and above commercial imports. Since 'usual' and 'normal' are difficult to define, most commentators agree that a partial distortion of traditional patterns of trade is inevitable under food aid with adverse consequences for commercial traders. Orderly disposal is intended to suppress the tendency for food prices to fluctuate or be depressed (disturbance), in the recipient country, caused by the release of food aid into the marketing system. Experience has shown that claims of a damaging impact on prices are often exaggerated owing to the relatively small volume of food involved. The third principle of consultation is designed to ensure that food aid is directed to approved sections of the population within the recipient countries, with suitable provision made for the marketing or disposal of food aid.

Cereals (27 per cent of food aid by value), SMP (39 per cent) and butter oil (35 per cent) are the main constituents of food aid from the EC. Whereas cereal shipments remained static during the 1970s, dairy products increased in importance reflecting the nature of agricultural surpluses in the EC rather than demand in the recipient countries. The composition of aid has fluctuated, however: in 1970, for example, 80 per cent of food aid was in dairy products, while four years later the figure was reduced to just over half, with an increase in cereals aid together with some sugar. Aid has tended to be concentrated in a few countries. Between 1969 and 1980 most aid was sent to Bangladesh, Pakistan, Egypt and India – significantly, all non-ACP countries. This concentration of aid was despite a three-fold system of identifying countries with a need for aid based on a cereals import requirement, an annual per capita GDP of less than $325, and an adverse balance of payments. The tight control of the Council of Ministers in approving food aid has allowed political considerations to influence the allocation of available aid. Consequently, there is no clear relationship between food need and the receipt of food aid.[30] Nevertheless, the EC's share of international food aid has risen from less than 1 per cent in 1965 to over 20 per cent by the late 1970s, with the share of cereals aid rising from 10 to 18 per cent over the same period.

Food aid from the EC, as from other developed countries, remains a contentious issue and a policy dilemma. If the Community, with its agricultural surpluses, does not dispense food aid to needy countries, Member States could be accused of immoral behaviour. Such aid has the potential to ameliorate suffering and release resources for economic development. If surpluses are given as aid, however, there can be long-run damaging consequences for the agricultural sectors of the recipient countries. Prices on domestic markets can be undermined, domestic producers deprived of the incentive to produce indigenous food crops, and agricultural production orientated towards export crops. A careful balance has to be struck, therefore, between too much and too little aid. Indeed, many commentators argue that if developed countries wish to give development aid to the Third World, it would be better channelled directly through financial loans and grants rather than indirectly through food aid. Certainly, food aid will not solve the problems of world hunger in the long term; developing countries must be encouraged to feed themselves.

Conclusion

From this analysis, the EC market appears not to be as closed as is often presented, but is clearly more open to products which do not compete directly with the agricultural output of the Community. 'Protection' is given on a selective basis so as to secure food supplies and insulate the internal market against fluctuations in world prices. From a European standpoint these are not seen as unreasonable objectives, particularly when account is taken of projections of future patterns of trade in agricultural products where 'uncertainty' is the dominant feature.[31] Nevertheless, traditional suppliers of 'temperate' food products to markets in Western Europe have undoubtedly suffered in relative, if not absolute terms from the import substitution effects of the policies pursued by the Community.

Food aid policy, by contrast, emerges as an arm of domestic agricultural policy in the Community. The importance of dairy products in food aid reflects the state of surplus production rather than the needs of developing countries. Even so, there are limits to the amount of food aid that can be given as far as development trends in Third World countries are concerned. While criticism can certainly be levelled at the total amount of development aid given by the Member States, increasing the volume of food aid is not necessarily the most efficient means of transferring resources, even though it accounts for only 15 per cent of all development assistance. More attention could be given to the use of surpluses in the creation of strategic buffer food stocks to regularise world food markets,[32] while estimates of future demands for food aid (100 million tonnes of cereals by 1991) suggest a continuing role for EC food surpluses for decades to come. In these circumstances, the need is for a long-term programme of food aid, depending more on the projected demand for cereals rather than dairy products, compared with the present short-term policy based on the disposal of available surpluses. The use of dairy products in India in Operation Flood remains the exception rather than the rule.[33]

The aspect of agricultural trade that causes most justified concern is the export policy of the EC. Although world trade in temperate-zone agricultural products is a small and declining fraction of total trade, in political terms its importance has been growing. The export policy of the Community raises issues such as the stability of world markets, divergence from free trade principles, and export dumping.[34] The need for a coherent export policy has become pressing with the farming lobby developing the concept of a 'vocation to export'. Agricultural exports are seen as a means of relieving the Community of excess production together with the need for producers to revise the pattern and volume of their production.[35] Unless producer prices can be reduced nearer to world market levels, the subsidised export of agricultural products from the EC will continue to be a cause of international friction and a justified focus for criticism of the CAP.

Notes

1 M. Loseby and L. Venzi, 'The effects of MCAs on EC trade in agricultural commodities', *European Review of Agricultural Economics* V (1978), pp. 361–80.
2 W. Prewo, 'Trade interdependence and European integration', in L. Hurwitz (ed.), *Contemporary perspectives on European integration* (London, Aldwych Press, 1980), pp. 77–94; F. Knox, *The Common Market and world agriculture: trade patterns in temperate-zone foodstuffs* (New York, Praeger, 1972), pp. 11 and 30.
3 F. O. Grogan, *International trade in temperate-zone products* (Edinburgh, Oliver and Boyd, 1972), p. 29.
4 The trade statistics quoted in this chapter have been calculated by the author from relevant volumes of Eurostat, *Yearbook in Agricultural Statistics* (Brussels).
5 Economic Commission for Europe, *Agricultural trade in Europe*, Agricultural Trade Review 16 (New York, United Nations, 1979), p. 19.
6 G. M. Taber, *Patterns and prospects of Common Market trade* (London, Peter Owen, 1974), p. 17.
7 Y–S. Hu, *Europe under stress*, (London, Butterworths, 1981), p. 50.
8 Loseby and Venzi, *European Review of Agricultural Economics* V (1978), pp. 361–80.
9 S. Young, 'Trade creation and trade diversion in the EEC: a case study for milk', *Journal of Agricultural Economics* XXVI (1975), pp. 197–208.
10 P. Mishalani *et al*, 'The Pyramid of Privilege', in *EEC and the Third World: a survey, Volume I* (London, Hodder and Stoughton, 1981), pp. 60–82.
11 A. Tovias, *Tariff preferences in Mediterranean diplomacy*, Trade Policy Research Centre (London, Macmillan, 1977), p. 67.
12 K–H. Beissner, 'The impact of the EC's agricultural policy on its trade with developing countries', *Inter Economics* XVI (1981), pp. 55–60.
13 C. Cosgrove Twitchett, *A framework for development: the EEC and the ACP* (London, George, Allen and Unwin, 1981), p. 92.

14 Known as the 'fluctuation threshold', this is usually set at 3.5 (previously 2.0) per cent and 7.5 (previously 6.5) per cent of the reference level.

15 Finance, limited to 375m. EUA for 1975–80 and 555m. EUA for 1980–5, is drawn from the European Development Fund (EDF). The EDF is discussed in detail by C. Cosgrove Twitchett, *Europe and Africa: from association to partnership* (Farnborough, Saxon House, 1978), pp. 39–51.

16 Economic Research Service, *The European Community's Common Agricultural Policy – implications for US trade*, Foreign Agricultural Economic Report 55 (Washington, USDA, 1969), p. iv.

17 Strictly there are four types of stocks in the EC: 1: intervention purchases (cereals, sugar, olive oil, beef, butter, tobacco, SMP), 2: private storage aids (butter, certain cheeses, beef, pigmeat, wine), 3: withdrawn produce (fruit and vegetables), 4: compulsory minimum stocks (sugar).

18 G. P. Sampson and A. J. Yeats, 'An evaluation of the CAP as a barrier facing agricultural exports to the EEC', *American Journal of Agricultural Economics* LIX, pp. 99–106.

19 I. R. Bowler, *Government and agriculture: a spatial perspective* (London, Longman, 1979), p. 28.

20 B. Balassa, *European Economic Integration* (Oxford, North-Holland Publishing Company, 1975), pp. 116 and 322.

21 G. R. Lewthwaite, 'New Zealand milk on the map', *Annals of the Association of American Geographers* LXX (1980), pp. 475–91; J. Lodge, *The EC and New Zealand* (London, Frances Pinter, 1982).

22 H. R. Wagstaff, 'EEC food imports from the Third World and international responsibility in agricultural policy', *European Review of Agricultural Economics* II (1974/5), pp. 7–21.

23 T. M. Shaw, 'EEC–ACP interactions and images as redefinitions of Eur Africa', *Journal of Common Market Studies* XVIII (1979), pp. 135–58.

24 A. Hewitt and C. Stevens, 'The Second Lomé Convention', in *EEC and the Third World: a survey. Volume I* (London, Hodder and Stoughton, 1981), pp. 30–59.

25 D. Jones, *Europe's chosen few: policy and practice of the EEC aid programme* (London, Overseas Development Institute, 1973).

26 R. Pasca, 'Mediterranean agricultural trade problems and the effects of the EC policies', *European Review of Agricultural Economics* V (1978), pp. 221–54.

27 R. B. Talbot, 'The EC's food aid programme', *Food Policy* IV (1979), pp. 269–84.

28 This volume was divided between the EC (720,500 tonnes) and the Member States (566,500 tonnes). The constitutional bases for EC food aid lie in Articles 43, 113 and 228 of the Rome Treaty.

29 J. Cathie, *The political economy of food aid* (Aldershot, Gower, 1982), p. 68.

30 J. R. Tarrant, 'EEC food aid', *Applied Geography* II (1982), pp. 127–41.

31 M. Tracy and I. Hodac (eds.), *Prospects for agriculture in the EEC* (Bruges, De Tempel, 1979), pp. 28–54.

32 T. Josling, *Developed-country agricultural policies and developing-country supplies: the case of wheat*, Research Report 14 (Washington D.C., IFPRI, 1980), p. 48.

33 Dairy products, especially SMP aid, from the EC are being used by the WFP to fund the development of dairy producer co-operatives in rural villages and distribution networks for milk and dairy products in urban areas.

34 S. Tangerman, 'Agricultural trade relations between the EEC and temperate food exporting countries,' *European Review of Agricultural Economics* V (1978), pp. 201–19.

35 M. Tracy, *Agriculture in Western Europe: challenge and response 1880–1980* (2nd ed.), (London, Granada, 1982), p. 368.

10
Changes in the farm-size structure of agriculture

Farm-size structure is arguably the central element in the problems faced by the CAP. Small farms predominate in the EC and agricultural prices have been set at levels sufficient to yield their occupiers a 'fair standard of living' (Article 39 (1B) of the Rome Treaty). Unfortunately, CAP price levels have been so favourable for the occupiers of large farms as to stimulate unwanted production. The resulting agricultural surpluses have precipitated the budgetary problems of the EC discussed in Chapters 4 and 5. Moreover, the price reductions needed to bring supply and demand into equilibrium, for products such as milk and cereals, would be so damaging socially for large numbers of farm families that to date they have been unacceptable to most governments in Western Europe.

The farm-size problem exposes a major weakness of the CAP, namely its over-reliance on price policy at the expense of structural policy. The latter, associated with the 'resource adjustment solution' to the farm problem, aims to reduce the number of small farms and increase the average size of those that remain through farm amalgamation. The imbalance of the CAP is evident in the small and declining proportion of the Agriculture Fund allocated to the Guidance Section (Table 7), which in part reflects the predominant role left to each Member State in matters of structural policy. The CAP has avoided contentious issues such as the ownership of the factors of production (land, capital), the preferred rate of decline in the size of the farm population, and the financing of the process of transition by the agricultural sector to a modernised structure of fewer but larger farms. These politically sensitive matters have been left to the Member States; the Community has merely provided a framework of Directives and partial funding for national programmes which promote farm-size change (see pp. 54–55).

The production trends discussed in Chapters 6, 7 and 8 have taken place in the context of, and been shaped by, national and regional variations in farm-size structure. Also, changes in farm-size structure have been part of the process whereby each agricultural product has become concentrated on fewer farms (see Table 36). In this chapter, therefore, attention is turned to the changing distribution of farm size within the Community and the relationships with farm incomes and wider social change in rural society.

Farm-size structure

The existence of marked variations in farm-size structure amongst the Member States has been suggested already in Tables 24–34. The United Kingdom and, to a lesser extent, Denmark display the largest average sizes of farm enterprise whether measured as the area of cereals (Table 24) or the number of dairy cows (Table 30). Similarly, Italy consistently records the smallest enterprise sizes, with other countries occupying variable positions between these extremes depending on the farm enterprise under review. Measuring the area-size of farms directly produces a ranking of countries with the United Kingdom, Luxembourg and France having the largest average sizes of farm, while Italy, West Germany and Belgium have the smallest (Table 46). Of farms in the EC, 79 per cent are still smaller than twenty hectares in size, but each country has a distinctive farm-size distribution. The United Kingdom, for example, has a particularly high proportion of farms greater than fifty hectares in size; Italy (like Greece) is dominated by holdings of less than five hectares; while Belgium and the Netherlands have a higher proportion of farms in the range five to ten hectares than other countries. These statistics should be interpreted with caution because they depend critically on how a farm or holding is defined,[1] and on what minimum size of farm enters the agricultural census. Definitions and practices vary from country to country. Thus exact comparisons of the different size categories should be approached with caution.

Farm sizes also vary considerably within each Member State and reflect regional differences in urban proximity, the physical environment, inheritance laws and the history of land settlement.[2] Looking first at the distribution of farms smaller than twenty hectares in size (Figure 14A), they predominate throughout Italy (Liguria, Campania, Abruzzi, Trentino), Belgium, the Netherlands, central parts of West Germany (Hessen, Rheinland-Pfalz), and southern regions of France (Roussillon to Côte D'Azur). Consequently, Italy contains nearly 40 per cent of all farms in the EC (over one hectare in size), and France a further 25 per cent. Localised

Table 46. Farm-size structure of the Member States, 1980ᵃ (% farms in each country)

Country	1–4.99 ha	5–19.99 ha	20–49.99 ha	over 50 ha	Average size (ha)	Annual rate of change 1970–6 in number of farms
United Kingdom	14(19)	28(29)	27(26)	31(27)	68.70	−1.7
Luxembourg	19(21)	28(37)	41(38)	13(4)	27.63	−4.2
France	20(23)	37(43)	31(26)	13(8)	25.41	−3.2
Denmark	12(12)	46(52)	34(31)	9(6)	24.96	−2.0
Ireland	15(20)	47(52)	30(22)	9(6)	22.52	−0.5
Netherlands	24(33)	51(51)	22(15)	3(1)	15.61	−2.5
Belgium	29(34)	48(51)	19(13)	4(2)	15.43	−4.0
West Germany	34(37)	43(46)	21(15)	3(2)	15.27	−3.2
Italy	69(68)	26(26)	4(4)	2(2)	7.42	−1.8
(Greece)	71(−)	27(−)	2(−)	0.2(−)	4.27	−
The Ten (The Nine)	46(43)	33(39)	15(15)	6(5)	18.0	−

a: there are rounding errors in the data; 1970 in brackets
Source: Barclays Bank, *Finance for farmers and growers 1982/83* (London, Barclays Bank, 1982), p. 46

areas with small farms occur in other countries such as the crofts of the Highlands and Islands of Scotland.

An alternative view of farm size can be derived from the amount of work provided by each holding. Standardised measures of labour input can be applied to each type of crop and livestock (annual work units, AWUs). This allows small but intensively farmed holdings to be separated from those that create under-employment for their occupiers. Figure 14B shows the proportion of holdings yielding less than one AWU, that is less than full employment for one person. As expected, there is a strong correlation between the distribution of these, and that of the smaller farms. The problem holdings are most numerous throughout Italy (60 per cent of all holdings), especially in the south. Elsewhere, Saarland, Hessen and Baden-Württemberg, Rousillon and Languedoc are problem regions within West Germany and France respectively. On the other hand, the small farms of the Netherlands and Tuscany, Emilia-Romagna and Veneto in Italy are shown to be less problematic than when judged by area-size alone.

Medium-sized farms in the context of the EC are those between ten and

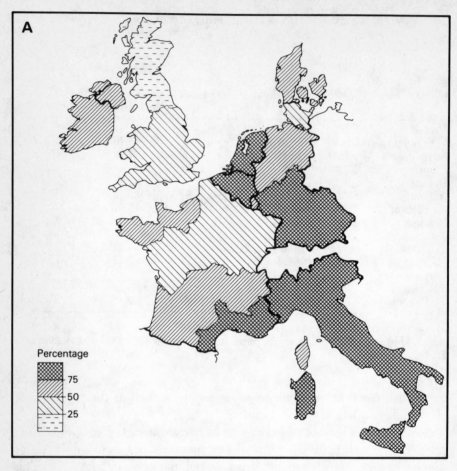

Figure 14A. Percentage of farms of less than twenty hectares, 1975

twenty hectares in size. They are characteristic of south-west France, Brittany, Nord Pas-de-Calais and Franche-Comté, and scattered parts of West Germany. This distribution of large farms, on the other hand, tends to be the inverse of that of small farms. Holdings over fifty hectares in size, for example, comprise more than a third of farm units in eastern, central and southern Scotland, northern, central and south-east England, and north-central France (Centre, Burgundy, Picardy, Champagne-Ardenne). South-east Ireland, Schleswig-Holstein in West Germany, and most parts of Denmark also have significant proportions of large farms.

One consequence of the 'structural imbalance'[3] between the numbers of

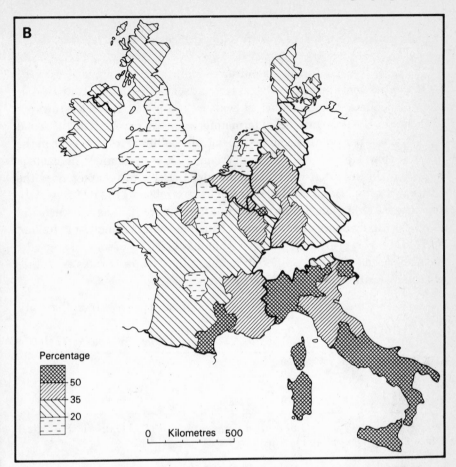

Figure 14B. Percentage of holdings with less than one annual work unit, 1975

small and large farms in a region is the degree of concentration of farm land in a relatively few but large holdings. Paradoxically, the degree of concentration (as measured by the Gini Coefficient[4]) is greatest in the two countries most dissimilar in farm-size structure – Italy and the United Kingdom. In Italy, 2 per cent of the largest farms account for a third of the farmland. The degree of concentration is highest in the south (Corsica, Basilicata, Puglia), centre (Lazio) and extreme north (Trentino-Alto Adige, Val d'Aosta) of the country. In the United Kingdom, 19 per cent of the largest farms cover over two-thirds of all farmland, a feature especially of the Highlands of Scotland and south-east England. Southern France is the only other extensive area where land is so concentrated, but even in Brittany, with its family-farm structure, the 34 per cent of holdings over twenty hectares cover 63 per cent of the agricultural land.

Of course, the structure of agricultural holdings was changing prior to the CAP and has continued to evolve in recent decades. Although statistics are often inadequate, there is sufficient evidence to suggest an accelerating rate of decline in the total number of farm units in the EC through the 1950s and 1960s (up to 4 per cent in some years), but with falling rates of decline in the 1970s (below 2.5 per cent in most years).[5] Average annual percentage reductions in the range 0.5 (Denmark) to 2.4 (Belgium) characterised the 1950s, for example, and 0.4 (Ireland) to 3.3 (Luxembourg) the 1960s. Overall, the greatest reductions in the number of farms since 1950 have occurred in Belgium, Luxembourg and the United Kingdom, and the smallest in Ireland and France. However, these changes, while cumulatively of some magnitude, have not been sufficiently different amongst the Member States as to alter national variations in farm-size structure.

Table 47. Annual change in number of agricultural holdings – West Germany (%)

Years	1–4.99 ha	5–9.99 ha	10–19.99 ha	20–49.99 ha	50+ ha	Total
1967–70	−5.9	−5.0	−2.5	+3.8	+3.6	−3.5
1970–3	−5.3	−5.7	−4.8	+3.3	+7.2	−3.7
1973–6	−4.0	−3.8	−3.7	+0.7	+4.5	−2.8
1976–8	−3.9	−3.6	−2.6	0	+3.6	−2.6

Source: Eurostat, Yearbook of Agricultural Statistics 1975–8 (Brussels, Statistical Office of the European Communities, 1980), p. 60

The process of farm-size change has been similar in most countries. The occupiers of the smallest farms have left farming in the largest numbers, and all farm-size categories below twenty hectares have declined in importance. In contrast, farms over twenty hectares have increased in relative significance, especially those over fifty hectares in size, as they have absorbed the land of farmers quitting agriculture. The trends for West Germany (Table 47) are typical for most countries in the EC. The economic threshold of farm size has been moving upwards in recent years, however, but at a pace limited by the favourable price régime of the CAP; the rate of structural change has been kept to socially acceptable levels by agricultural policies in the Community.[6] In most countries in the 1950s, for example, the proportion of farms in the size range ten to twenty hectares was increasing. By the 1960s this group was in decline, and by the end of the 1970s there is some evidence of falling numbers starting to effect farms in the range twenty to fifty hectares. This underlines the long-term and continuing process of structural change in agriculture whereby the productive resources of the smallest, least efficient farms are absorbed by larger farming units. The threshold of economic viability tends to move inexorably upwards, and only in a few regions in France and the United Kingdom have the contemporary upper limits to the process apparently been reached. Recent research has shown that after a certain size, usually sufficient to give full-time employment to two or three workers, there are no economies of scale to be achieved.[7] Resources are not used less efficiently on very large farms, but neither are they used more efficiently. Most regions of the EC, however, are far away from this upper economic limit to the process of farm-size change.

Thus quite spectacular reductions in the number of farms have occurred in the Member States over the last three decades. Unfortunately, the national rates of decline have never been sufficient to resolve the farm income problems of those who have remained in production. Consequently, in the 1950s and 1960s all countries developed programmes to help speed up the process of withdrawal from agriculture. Although they varied in detail, most contained three elements. First, retirement pensions or grants were made available to farmers wishing to give up farming and sell their land for amalgamation with other holdings. Finance was usually restricted to older farmers and given in proportion to the amount of land released. Secondly, grant aid was commonly given to those farmers who wished to expand their farm businesses either by purchasing or renting available land, or by intensifying production from their existing holdings. Thirdly, some countries established either a national authority, or a set of regional agencies, charged with regulating the process of farm amalgamation often by entering the land market to purchase and dispose of available land. The regional *Sociétés d'Aménagement Foncier et d'Établissement Rural* (SAFERs) of France, the Land Commission of Ireland, and the *Stichting Beheer Landbouwgronden* of the Netherlands are perhaps the best known authorities involved with farm amalgamation. Since the various national programmes of structural change have been extensively reported elsewhere,[8] and the concern here is with the CAP, attention is now focussed on Community initiatives in the context of farm-size change.

The 'Mansholt Plan' of 1968 (see Chapter 3) turned attention in the Community towards the need for farm-size change and a reduction in the agricultural work force. The politically powerful farming lobby ensured that the far-sighted proposals of that Plan were delayed and diluted. When

Directives emerged in 1972 they reflected existing policy measures in the Member States.

Directive 72/160/EEC enabled Member States to provide or continue national retirement pensions for farmers leaving agriculture. Generally, national exchequers are refunded 25 per cent of the costs involved from the Guidance Section of FEOGA. In the United Kingdom, for example, under a 'Payments to Outgoers' scheme that can be traced back to the 1967 Agriculture Act, payments are made to owner-occupiers and tenants who give up their uncommercial holdings for amalgamation with farms having an approved 'development plan'. Recipients of aid must be between fifty-five and sixty-five years of age and be full-time farmers. They receive either a lump sum payment on a sliding scale according to the amount of land released (to a maximum of £3,000),[9] or an annuity. Farmers not qualifying by reason of age or part-time farming can receive payments at a lower rate.

Experience in most countries suggests that the level of payments has not been sufficient to alter materially the rate of out-migration by farm occupiers. More likely, pensions have been taken by farmers who would have left agriculture in any event. The number of farmers applying for a retirement pension, and thus freeing land for amalgamation, has been falling since 1975. The protection of agricultural incomes by the CAP, the exclusion of part-time farmers, and diminishing opportunities for off-farm employment are additional reasons for the poor response to Directive 72/160 (Table 49). In the six years between 1975 and 1981, for example, only 77,042 farmers throughout the EC took a retirement pension; 1,045,000 hectares of land were released, representing 1.1 per cent of the agricultural area of the EC and an average of fourteen hectares per recipient. In return, 96,964 holdings were increased in area by an average of 6.7 hectares, although only fifteen per cent of such holdings had an approved development plan. Nevertheless, these farms acquired larger blocs of land compared with farms not having a plan. Most of the land and the recipients of aid have been located in France (east and south east) and West Germany (north), although in relative terms the impact has been greatest in Luxembourg. The rate of uptake of retirement grants has varied being greater near urban areas where alternative employment opportunities are more frequent. In most countries farms of less than ten hectares have been released for amalgamation; only in Ireland and the United Kingdom have holdings greater than twenty hectares in size been greatly affected. Once again these differences reflect underlying variations in the farm-size structure of each country.

Other reasons can be advanced for the failure of structural measures to speed up the rate of farm-size change. These include the financial and emotional costs of geographical and occupational mobility for the individual, the elderly age-structure of the farm population, the absence of skills needed in urban-industrial employment, and a range of 'survival strategies' that can be adopted by the occupiers of small holdings. Where resources permit, farmers can acquire more land or intensify their production. Alternatively, members of the family can obtain off-farm employment while helping to maintain the farm, the occupier becoming a part-time farmer. A further strategy is to form a co-operative with other farmers to either purchase farm inputs, market farm produce, or combine in the production of a particular product. Known respectively as supply, marketing, and production co-operatives, they can be operated as primary co-operatives or in regional federations as secondary or even tertiary-level co-operatives. Co-operation allows farmers to reduce their production costs or gain higher product prices by negotiating with farm suppliers or marketing agents (countervailing power). In addition, co-operatives generate economies of scale and integrate the local economy with the national market. Indeed, attention has been drawn already to situations where a strong regional network of co-operatives can create an 'agglomeration economy' and so strengthen an area's product specialisation.[10] In the Midi, for example, wine co-operatives have achieved a ten per cent price premium over individual producers to the advantage of the small farm.[11]

Most countries provide technical and financial assistance to farmers who wish to set up a co-operative. In fact, under the CAP grant aid can be provided for capital projects concerned with the improvement of marketing and processing in agriculture (Regulation 355/77), the setting up of fruit and vegetable producer organisations (Regulation 1035/72), and the setting up and management of production groups (Directive 72/159), including forage groups in the LFAs (Directive 75/268). Nevertheless, the development of co-operatives is very variable within the EC both by product and location. Relatively few co-operatives, for example, are found in Italy, Belgium and the United Kingdom (Table 48). In these countries fewer than one quarter of all farmers are members of a co-operative. In the United Kingdom the heterogeneity of production patterns and the operation of government marketing boards have inhibited the development of co-operation, while in Italy the predominance of very small farm units has been the hindrance. Well-developed co-operative structures, by comparison, occur in France, Denmark and the Netherlands. In these countries finance for development, training schemes for managers, and the evolution of secondary and tertiary tiers have all favoured co-operatives.[12] In France, for instance, secondary-tier co-operatives have developed for wine, milk/dairy products, and farm requisites in the Languedoc, Basse-Normandy/Pays de la Loire, and northern France respectively.[13] With the concentration of co-operatives, however, the number of primary groups is now falling. In Denmark there was a 42 per cent loss of primary co-operatives between 1960 and 1978, and in West Germany a loss of 36 per cent in milk co-operatives alone between 1971 and 1978.[14]

Table 48. Co-operative market shares for selected product markets, 1977 (% national market)

Product	Country[a]								
	F	N	D	WG	L	Ir	It	B	UK
Milk	48	87	87	78	90	88	35	65	–
Cereals	67	60	50	52	90	23	15	15	17
Vegetables	30	84	70	36	–	22	5	55	10
Fruit	40	10	65	26	30	14	60	40	19
Eggs	25	20	59	–	90	25	5	–	19
Pigs	52	26	90	20	–	25	3	15	7
Poultry	42	32	50	–	–	50	10	–	2
Beef	19	18	60	19	22	33	5	–	6
Sugar beet	17	61	14	–	–	–	15	–	–

a: country abbreviations as on Table 39
Source: G. Foxall, *Co-operative marketing in European agriculture* (London, Gower, 1982), p. 2

Parallel with, although often separate from, farm amalgamation has been the process of farm consolidation. In several Member States, the land comprising individual farms is often fragmented into small plots and widely dispersed around the residence of the farmer. The processes causing land fragmentation are many and varied including land inheritance, population pressure, local topography, fossilisation of open-field systems of farming, equal division of different soil types, and risk-minimising in environmentally hazardous localities.[15] Many governments have funded programmes to consolidate the scattered fields of each owner and so rationalise the pattern of land ownership. France, Denmark, West Germany, Italy, Ireland and the Netherlands have all had programmes of land consolidation, each meeting with varied success. Not only has progress been slow and spatially uneven, but programmes have been costly and the size of farm created has often been too small for economic viability. Land consolidation, including new roads, drainage systems and resettlement away from congested villages, therefore, has often been a necessary prelude to farm amalgamation. Since land consolidation remains a national responsibility under the CAP, it is not considered further here. Rather the reader should refer to one of the many summaries of the topic which outlines the economic consequences of a fragmented pattern of land ownership and the various national programmes of land consolidation.[16]

Lastly in this section, one of the ironies of the contemporary process of farm amalgamation is that a fragmented pattern of land occupation is being recreated, albeit at a farm rather than field scale. Farmland comes onto the market in a somewhat haphazard way as individuals either die, retire or decide to leave agriculture. Even where land agencies operate, the market economy is allowed to function so that the highest bidder commonly acquires the available land. The amalgamator farm and the newly acquired farmland are often separated by a considerable distance so that a new pattern of farm fragmentation emerges.

Farm incomes
Disequilibrium between supply and demand for agricultural products tends to exert a long-term, downward pressure on farm incomes. The comparison of net farm with non-farm incomes, however, has been subject to a great deal of methodological debate. There are national variations in accountancy conventions on matters such as the separation of farm from household expenses (notional rent for farmhouse, food produced and consumed on the farm), allowances for the increasing value of farmland, opening and closing valuation of assets, allowances for the manual and managerial input of the farmer, allowances for the labour input of the farm family. Choosing an appropriate non-farm group for a comparison of incomes is also problematic owing to the wide range of functions undertaken by the farmer – provider of capital, manager, and manual worker. Moreover, the farms in each national accountancy data network are seldom representative of agriculture, most being large in business-size and commercially-oriented. Generally, therefore, farm and non-farm incomes tend to be treated in aggregate, while for agricultural comparisons 'gross value added per annual work unit' (GVA/AWU) is widely used as a surrogate of farm income.

Bearing these problems in mind, and excluding Belgium, the Netherlands and the United Kingdom,[17] farm incomes in aggregate tend to fall below non-farm incomes in most years. Occupiers of the small farms do not possess sufficient resources of capital and land to generate incomes comparable with those earned elsewhere in the economy. Clearly, the situation can change

through time. GVA/AWU rose consistently between 1968 and 1973 in the
Community, declined between 1973 and 1975, remained relatively stable
until 1978, and then declined further. During the same period, real incomes
per head in the Community economy moved upwards, with a less rapid rate
of increase between 1973 and 1975. The most dramatic increases in
GVA/AWU were recorded in France for the years 1969 to 1973, and in Ireland
and Italy between 1972 and 1978; the most severe downturn in farm incomes
occurred in the United Kingdom after 1973; other countries display relatively
stable farm income levels. Taking the period 1973 to 1981, agricultural
incomes in real terms increased in Italy (plus 3 per cent), Netherlands (plus 5
per cent), Belgium (plus 6 per cent) and Denmark (plus 8 per cent), but
decreased in all other countries especially France (minus 24 per cent), the
United Kingdom (minus 21 per cent) and West Germany (minus 22 per
cent).[18] One estimate attributes three-quarters of the change in GVA/AWU
to decreases in the farm labour force (structure), and the remaining quarter
to the increased productivity of resource use (intensification).[19]

Farm incomes vary with farm size. Large farms, because of the sheer
magnitude of the resources involved (increasing capital/labour ratio with
size) and the economies of scale that can be realised, yield their occupiers
incomes comparable with those earned elsewhere in the economy. This is
despite the returns per unit of capital, per hectare of land, or per labour unit
being inferior to those obtained on more intensively-worked small farms.
Moreover, the benefits of price support under the CAP accrue disproportionately to those larger farms which produce most of the agricultural output
of the EC. Farm incomes also vary with farm type not least because of the
differential price support given to various products under the CAP. In the
United Kingdom, for example, holding size of farm constant, cereal farms
have consistently out-performed livestock farms over the past two decades in
terms of farm income. Since both farm type and farm size vary from region to
region within the Community, not surprisingly farm income also exhibits a
strong spatial variation. In addition, transport cost advantages can be gained
by those farms located near to urban markets, while in peripheral regions,
often including mountain and upland areas, farms suffer increased production costs owing to location, climate and steepness of land. Where disadvantages in farm size and farm type are assocated with a peripheral location,
income levels can be at their lowest.

On a Community-wide basis, farm incomes are higher in the north, compared with southern regions, especially in a sweep of regions from central
France to Denmark, including eastern England and Scotland (Figure 15A).

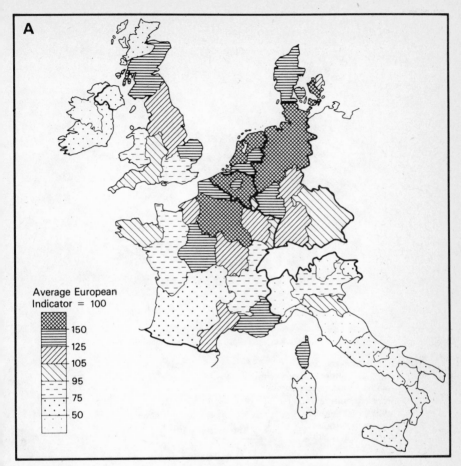

Figure 15A. Gross added value per annual work unit, 1976/7 (farm income)

These areas produce the cereals, milk and sugar beet which attract most of the direct financial support of FEOGA (Figure 7A). A similar association is revealed when account is taken of indirect levels of price support through limitations on food imports from outside the EC.[20] Northern regions also tend to have relatively high outputs of produce, both per hectare and per worker, and have increased production to the greatest extent under the CAP (Figure 7B). The high income regions include Champagne-Ardenne, Noord-Nederland, Niedersachsen, Nordrhein-Westfalen, Lorraine, Centre and Alsace. The low-income regions cover all of Italy, especially Puglia,

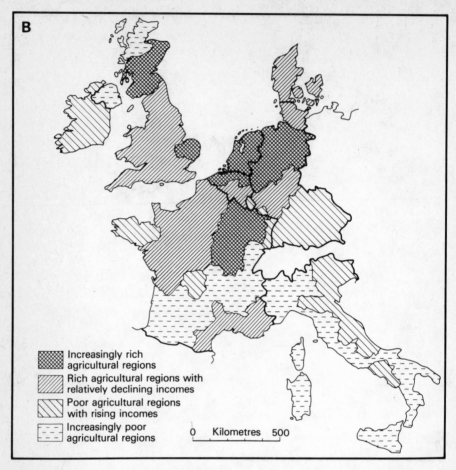

Figure 15B. Typology of regional trends in agricultural incomes, 1968/9–1976/7

Basilicata and Molise, Northern Ireland, Limousin-Auvergne, Val D'Aosta and south-west France.

One of the most severe criticisms levelled against the CAP is that despite the most complex of price support mechanisms, average real farm incomes in most countries have not been raised significantly relative to non-farm incomes, nor have regional variations in farm income levels been reduced. Indeed, the CAP has exacerbated rather than relieved regional income differences. Those areas with the lowest farm incomes in the early 1950s still have the lowest incomes today; conversely, the high-income regions have remained the same.[21] In Italy, for example, the ratio of minimum to maximum regional farm income was 1:3.2 (Basilicata: Lombardy) in 1965; by 1977 the comparable ratio was 1: 4.5 (Molise: Lombardy).[22] Increases in regional farm income disparities have been recorded elsewhere in the EC. In the United Kingdom, for example, the ratio of regional farm incomes deteriorated from 1: 1.52 (Northern Ireland: north/east/west England) in 1971, to 1: 1.79 (Wales: Scotland) in 1977. In France, the position has improved slightly from 1: 7.8 (Limousin: Ile de France) in 1970 to 1: 4.6 (Basse-Normandy: Ile de France) in 1977, but in West Germany the familiar pattern is repeated – from 1: 1.63 (Hessen: Schleswig-Holstein) in 1972/3 – to 1: 1.72 (Rheinland-Pfalz: Schleswig-Holstein) in 1976/7.[23]

Trends in farm income are summarised in Figure 15B using a fourfold typology. In northern regions, relatively declining average incomes in real terms are evident throughout England and Wales, west-central France, southern parts of Belgium, and Hessen/Rheinland Pfalz in West Germany. In southern parts of the EC, relatively rising incomes characterise the eastern seabord provinces of Italy from Veneto in the north to Abruzzi in the south, including Campania. Elsewhere in the north and south, income levels have continued to improve and deteriorate respectively.

Structural measures and farm incomes
The evolution of structural measures to aid farm incomes under the CAP has so far passed through four stages (see Chapter 4): the funding of individual projects to raise farm incomes, directives to modernise agriculture, direct income support to farmers in Less Favoured Areas, and regionally specific funds to aid investment in Mediterranean areas of the EC.[24] Generally, Member States receive a quarter of the costs of implementing these measures from FEOGA, but higher rates apply in defined problem regions in Italy and Ireland.
Regulations 17/64 and 355/77
From 1964, grants have been made annually for 'individual projects' in agriculture concerned with rationalising or developing storage, market preparation and improving information on prices. Since 1977 grant aid has been limited to projects concerned with production and marketing, provided they again form part of comprehensive and coherent sectoral or regional development projects. Forestry and fishery developments have also been eligible for aid. Projects can be promoted by public, semi-public or private bodies provided they are vetted and approved by the Member State before being forwarded to FEOGA for funding. Only substantial projects costing more than approximately £100,000 have been considered, while funding has

been on a competitive basis since applications always exceed the funds available. Three countries – West Germany, France and Italy – received over three-quarters of approved aid between 1964 and 1978, over a third being paid to West Germany. Unfortunately, the low rate of completion of projects has reduced the amount of aid actually paid to Italy compared with the number of approved projects. Under Regulation 355/77, Italy has again received most assistance (Table 50).

There is some evidence that aid has been channelled to problem regions in each country. In France, for example, payments have been steered towards the western (24 per cent of aid between 1964 and 1978) and the Mediterranean regions (18 per cent). The disbursement of aid by product reflects the agricultural systems of these regions – 37 per cent on wine, 23 per cent on fruit and vegetables and 22 per cent on meat.[25] In the United Kingdom, a large proportion of the £56.6 million worth of aid between 1973 and 1978 went to projects in Scotland (25 per cent) and Northern Ireland (21 per cent), with emphasis on meat, cereals, milk and fruit/vegetables. A flavour of the type of agricultural project receiving aid can be gained from the following list taken at random from the 47 successful projects in the United Kingdom in 1981 (£8 million total investment):

Construction of a cereals processing plant (Devon)	£329,110
Improvement of a vegetable packing plant (Lincolnshire)	£40,300
Construction of a milk processing plant (Norfolk)	£214,470
Construction of a potato store (Humberside)	£52,928
Construction of a slaughterhouse (Dyfed)	£136,835
Improvement of a milk processing plant (Aberdeen)	£86,972
Construction of a pigmeat processing factory (Co. Antrim)	£199,724

Directive 72/159/EEC

The Directive assists farm modernisation selectively on potentially viable farms. Financial aid passes to individual farms either as interest rate subsidies or capital grants. National schemes vary in detail so attention is given here to the Agriculture and Horticulture Grants Scheme (AHGS) as operated in the United Kingdom since 1980. Prior to that date, similar assistance had been provided under a Farm and Horticultural Development Scheme (FHDS). Financial aid is confined to farm businesses capable of producing a 'comparable income' for the occupier. This is taken to be the average income of comparable non-farm socio-economic groups in the same region, and in 1982 was set at £6,300 per full-time worker. The following items can be grant aided: permanent buildings, silos, water/electricity/gas supply, field drainage, roads, fences, grassland improvement, plant and machinery, and shelter belts. Rates of assistance are usually 32.5 per cent of total cost, but they vary with the item of investment and increase in value for farms in LFAs. Guidance premiums are also available to assist in the development of cattle and sheep production. To qualify for aid the farm must have an approved 'development plan' designed to maintain or raise the farm income to the 'comparable income' over a period of six years. In addition at least half the income of the recipient must be derived from agriculture, at least half of the recipient's working time must be devoted to the farm business, and the recipient must have at least five year's experience in farming or an appropriate vocational qualification. Limits to the AHGS are a minimum of £1000 per farm plan and a maximum of £46,500 per labour unit, with profitable farms excluded unless they are judged to be 'at risk'. Limits are also placed on aid for intensive pig units, while between 1977 and 1980 aid was withdrawn from dairy farms in an attempt to curb the rise in milk production.

Most studies concerned with the impact of Directive 72/159 have emphasised the low rate at which development plans have come forward from agriculture relative to the total number of eligible farms (only 172,779 by 1981). For example, fewer than 6 per cent of farms larger than ten hectares in size had an approved plan by 1978. By 1980, West Germany (28 per cent) and the United Kingdom (17 per cent) accounted for the greatest numbers of approved plans, although the highest proportional response rates from eligible farmers occurred in Denmark and the Netherlands, and the lowest in France. The volume of investment per agricultural worker has been high in West Germany and the Netherlands (mainly over 40,000 ECUs), but relatively low in Ireland (mainly less than 20,000 ECUs). Taking response and investment rates together, most investment under Directive 72/159 has taken place in West Germany (35 per cent) and the United Kingdom (29 per cent of all EC grant aid).

With the onus placed on individual farmers to apply for aid, there has been a variable pattern of response within each country.[26] In West Germany, response rates have been highest in Schleswig-Holstein, while in France the adoption of grant aid has been greatest in the Paris basin, Champagne-Ardenne, Brittany and Auvergne, and least in Basse-Normandie, Centre and Côte D'Azur. In the Netherlands and Belgium, areas of horticultural production have received most financial assistance. Overall, aid has gone to regions with large farms (twenty to fifty hectares) and cattle production, especially dairying. The majority of problem farms in the small size group, however, remain untouched, while only a quarter of farms with development plans have increased their area size. This last aspect reflects the failure of Directive 72/160 as discussed previously, but has lead most grant-aided farms to intensify their production and so contribute to the over-supply of agricultural produce. Most seriously, one analysis has shown that 58 per cent of farms with approved grant aid will not reach even 80 per cent of the 'comparable income' after six years with an approved development plan.[27]

Directive 72/161/EEC

This Directive promotes the training of farmers and farm workers for employment within agriculture, including the setting up of a network of socio-economic advisers within each Member State to give the farming community a better understanding of employment opportunities both within and outside agriculture. National responses have varied. West Germany and the Netherlands initially employed considerable numbers of advisers; the United Kingdom appointed only one adviser to each of the major administrative regions of the country; Italy, Ireland and Luxembourg at the outset had no structure of agricultural advisers. Most funding under the Directive has gone to France and West Germany for schemes of basic training for young farmers.

Directive 75/268/EEC

The objectives of the Directive are to safeguard agricultural activities in regions with permanent natural handicaps to production, maintain the density of the rural population, and conserve and manage the landscape. These objectives are reached by providing direct income payments (transfers) to producers in the form of Compensatory Allowances and by restricting payments to farmers in defined Less Favoured Areas (LFAs) (see Chapter 4 and Figure 4). The Directive covers a quarter of the agricultural area of the Community, 15 per cent of farm units and 12 per cent of the EC's agricultural output. The details of the payments vary from country to country, but in the United Kingdom they are paid annually to occupiers of at least three hectares of eligible land on each head of cattle or sheep kept in regular breeding herds and flocks. Recipients must undertake to continue to farm for a further five years except when retiring or discontinuing agriculture for the purpose of structural improvement under Directive 72/160. Most countries have an upper limit to the number of livestock eligible for payments to discourage over-stocking, or else make payments per hectare of eligible land.

Over half a million farmers in the EC receive income payments each year under the Directive, the majority being in France, Italy, Ireland and West Germany (Table 49). The average allowance per hectare and livestock unit

Table 49. Guidance payments in the Less Favoured Areas

Country	Directive 72/159a				Directive 75/268 (in 1981)			
	% national agricultural area	% development plans in LFA	% LFA farms with planb	% non LFA farms with planb	Number of beneficiaries	Gross allowances (ECU)		
						per ha	per l.u.c	per beneficiary
Luxembourg	99	n.d.	n.d.	n.d.	3772	15.4	65.2	1612
Greece	70	n.d.	n.d.	n.d.	219620	13.0	75.0	259
Ireland	54	11	0.9	11.2	94756	13.5	46.8	572
Italyd	41	n.d.	n.d.	n.d.	123132	1.1	53.2	319
United Kingdom	41	29	4.7	3.1	43402	18.9	65.1	3038
France	35	33	0.3	0.3	139574	10.4	51.1	1000
West Germany	30	18	1.6	3.7	82495	11.3	40.7	541
Belgium	19	14	3.2	3.0	10045	11.4	41.2	744

a: by 1977; b: based on total farms; c: l.u. – livestock unit; d: From 1978 and for five regions only (Piemonte, Tuscany, Basilicata, Trentino, Abruzzi)

Source: Arkleton Trust, *Scheme of assistance to farmers in less favoured areas of the EEC*, (The Trust, Langholm, 1982), pp. 8–10; P. Henry, *Study of the regional impact of the Common Agricultural Policy*, Regional Policy Series 21, RICAP Working Group, (Brussels, Commission of the European Communities, 1981), p. 44; Commission of the European Committee, *The Agricultural Situation in the Community, 1983 Report*, (Brussels, 1984), p. 304

onsiderably from country to country, and when related to average farm size
are most generous in the United Kingdom and Luxembourg. However, while
a uniform rate of payment is made throughout the United Kingdom, other
countries vary payments regionally according to the severity of the farm
income problem. In Ireland and France, for example, higher rates of payment
are made on sheep in certain mountain areas. Higher rates of grant aid are
also paid under Directive 72/159 in LFAs. Except in France, the United
Kingdom and, to a lesser extent, Belgium such grant aid has had even less
impact than elsewhere in the agricultural sector. (Table 49).

Criticism of Directive 75/268 takes two forms. First, payments tend not to
be sufficiently selective as to meet the social objectives of the Directive. The
farm income problem varies within defined LFAs depending on the presence
of products such as cereals and milk which receive price supports by other
means. Thus, in relative terms the average incomes of farmers in the
following regions are not problematic even though they receive
Compensatory Allowances: Wales, south-west England, mid-west and
south-west Ireland, Franche Comté, Auvergne, Baden-Württemberg and
Bavaria.[28] Equally, large farms benefit disproportionately compared with
small farms when Compensatory Allowances are paid per head of livestock,
even though their income problems are less severe. Approximately 30 per
cent of all Allowances in the EC, for example, are taken by 10 per cent of
beneficiaries with the largest holdings.[29] In the United Kingdom, 4 per cent of
the largest farms receive 22 per cent of the payments.[30] Directive 25/268,
therefore, has been of marginal importance in redressing the problem of low
farm incomes and stemming population outmigration from Less Favoured
Areas.

A second criticism lies in the use of agriculture as a means of achieving the
environmental objectives of the Directive. The assumption that a prosperous
agriculture is the best means of safeguarding the landscape and the
environment is now increasingly questioned. Farmers in upland areas are
equally likely to intensify their production by ploughing rough-grazings,
clearing woodland and draining wet areas as producers in lowland regions.
Moreover, the view that waste land – land temporarily abandoned into social
fallow – has negative effects on landscape is increasingly challenged.[31]
Whichever view is taken, aid given indirectly through income payments is a
most uncertain method of ensuring the conservation of the environment of
upland areas.

The Guidance Section of FEOGA
Despite doubling in value since 1974/5, payments under the socio-economic

directives remain a small part of all expenditure under FEOGA. A 1962 Regulation providing that structural measures should receive up to one third of all expenditure on market support has never been fulfilled. Most of the Guidance Section has been expended on Directive 355/77, although more recently Directive 75/268 has assumed an increasing importance (Table 50). The allocation of expenditure under each directive varies nationally. In part, this stems from the delay in some countries, such as Italy and France, in applying certain directives. There have also been variations in applying the full range of measures under each directive, such as the decision of the United Kingdom not to provide grant aid for tourist developments on farms under Directive 75/268. National variations in farm-size structure, and the availability of competing national schemes of assistance, also account for varying expenditure between the Member States. Overall, Italy, West Germany and France have gained most from the Guidance Section (Table 50).

Nevertheless, the areas favoured by expenditure from the Guidance Section have been the agricultural problem regions of the EC: all of Italy including Sicily, southern France, all of Ireland including Northern Ireland and Scotland.[32] The extent to which this pattern of expenditure, when added to national schemes of support, compensates for the disadvantages created by the Guarantee Section of FEOGA is uncertain. Most observers, however, point up the increasing rather than decreasing regional disparities of income as evidence of failure to correct for the unequal benefits that accrue to regions producing milk, cereals, and intensive livestock under the CAP.

The farm community

Thus far, attention has been focussed on the farm structure of the EC. The farm community, however, is comprised of both farm occupiers and farm workers. Another 'survival' mechanism open to farmers has been a reduction in their production costs by replacing expensive farm labour, both hired and family labour, by cheaper capital in the form of machinery. In a sense, the migration of the young to the towns in large numbers has been necessary for the survival of the family-farm structure of Western Europe. In general, the rate of decline in the number of farm workers has been greater than that of farm occupiers, although with the economic recession of recent years both rates of decline have fallen. By 1980 farmers and farm workers together comprised only 8 per cent of the workforce of the EC, whereas twenty-five years earlier the proportion had been 19 per cent. It is worth emphasising the 46 per cent reduction of the agricultural workforce of The Six between 1960

Table 50. EC expenditure on common socio-structural measures, 1974–9 (% in each country)

Country	Marketing & processing (Reg. 355/77)	Less Favoured Areas (Dir. 75/268)	Policy measure			Total expenditure
			Farm modernisation (Dir. 72/159)	Vocational training (Dir. 72/161)	Cessation of farming[a] (Dir. 72/160)	
Italy	41	2	1	7	–(–)	27
West Germany	14	15	35	19	82(10)	25
France	18	26	4	62	3(7)	24
United Kingdom	8	38	29	3	4(2)	13
Netherlands	5	–	13	0.1	1(2)	7
Ireland	6	15	8	4	8(1)	6
Belgium	3	3	1	4	2(3)	5
Denmark	14	–	9	1	–(–)	3
Luxembourg	0.2	0.8	–	–	–(13)	0.2
(% all EC expenditure)[b]	(45)	(29)	(25)	(1)	(0.2)	–

a: in brackets, recipients by 1977 as % farmers aged 55–65; b: average of 1978 and 1979

Source: M. Tracy, *Agriculture in Western Europe*, (Granada, London, 1982), p. 347; European Communities, *Grants and Loans from the European Community*, Periodical 7–8/1981, (Brussels, European Documentation, 1981), p. 15

and 1973 and the further 19 per cent reduction in The Nine from 1973 to 1980. By maintaining farm incomes at a higher level than they might otherwise have attained, however, the CAP has probably slowed down the rate of outmigration from agriculture.

The greatest proportional losses of farm employment have occurred in Italy, Ireland and France, although there have been regional variations within each country. In France, for example, rural depopulation has been most severe in central parts of the country, in Limousin-Auvergne and adjacent areas of Midi-Pyrénées, Languedoc and Rhône-Alpes.[33] Rural, non-farm population increases, by contrast, have recently been recorded near the industrial areas of the Paris basin, Lorraine, Alsace, Franche-Comté, and the Mediterranean *départements*. The United Kingdom and the Netherlands, already with relatively small agricultural workforces in the late 1950s, have experienced relatively low rates of decline in farm employment (Table 1). Only Ireland, Italy (mainly in the south) and Greece of the Member States still have sizeable proportions of their workforces in agriculture, although some regional concentrations are evident in France (Midi-Pyrénées, Auvergne, Normandy, Brittany – Figure 16A).

The transformation of rural society which has accompanied the decline of farm employment and the development of a market-oriented agriculture is a recent experience in most parts of the Community. Two broad social changes in agriculture have accompanied the modernisation of farming.[34] First, the selective rural outmigration of the young has left an ageing agricultural workforce. Only 23 per cent of those employed in agriculture are under thirty-five years of age compared with 41 per cent in the total population. The problems of an ageing workforce are most severe in Ireland, Luxembourg and Denmark where over a third of workers in agriculture are over fifty-five years of age (EC – 25 per cent). When an old-age structure of farmers is combined with small farm-size structure, the problems of 'structural-rigidity' in farming are compounded. Nor can young people easily find an entry into farming owing to the high cost of farmland and the falling number of farms in agriculture. In addition the structure of farm labour has been affected by farm-size changes and rural outmigration. More part-time, spare-time and seasonal labour is employed in the place of full-time workers, and as a consequence the workforce now contains a higher proportion of female workers. The implications of the 'feminisation' of the workforce have yet to be fully researched. On the one hand female workers receive wage-rates inferior to the full-time male employees they have replaced; on the other hand more wives of farmers are involved in decision making on the farm,

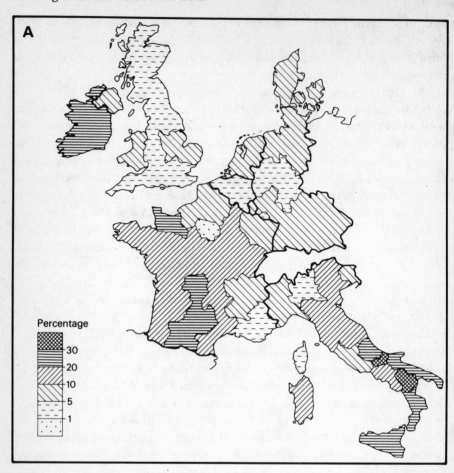

Figure 16A. Percentage of the working population employed principally in agriculture, 1977

especially where new forms of income are concerned – for example from tourism or direct farm sales.

The second broad consequence of farm-size change and rural outmigration has been the development of a new social order amongst farm families.[35] The fullest expression of the modernisation of agriculture is represented by the large, capitalist farm, often managed rather than owner-occupied, employing hired labour. This economically efficient stratum of farms produces a disproportionately large volume of the agricultural output of the EC. Such farms are often viewed as an extension of agribusiness capitalism

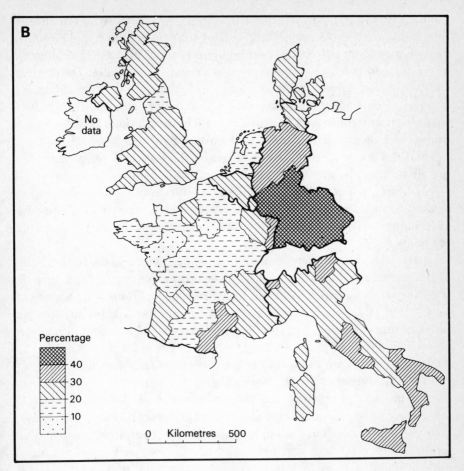

Figure 16B. Percentage of farmers having another gainful activity, 1977

so closely are they tied to industrial firms producing the inputs to agriculture (cattle feed, fertilisers), the food processing industries and non-farm capital such as banks, pension funds and investment companies.

A second stratum is comprised by full-time farmers on family-farms run on modernised, capitalistic lines. The persistence of family farms has been remarked on by many writers[36] and the interests of their occupiers are defended by the CAP. Amongst the stated advantages of family farms are the protection of rural value systems, the continuation of public services in rural areas, the maintenance of the ecological balance in the countryside, and the avoidance of over-concentration of population in congested areas. The profitability of family farms is threatened by growing indebtedness following the purchase of land, machinery or new buildings, while more radical observers point up the control over input and output prices and production methods than can be exercised by agribusiness monopolies. Dairies, for example, often appear to operate as local monopolies, with producers irreversibly dependent on them for the sale of their milk because of product specialisation.[37] Since food processors obtain a rising proportion of final food prices, such observations cannot be dismissed lightly.

The third social stratum is comprised by the occupiers of economically marginal farms. Some writers make a distinction between large and small peasant farms where 'peasant' describes a condition and set of values rather than an occupation. Such farms are seen as a survival of pre-capitalist modes of production being worked by non-profit-oriented, usually ageing occupiers.[38]

Lastly, part-time farmers form a dynamic social group in the countryside. Economists have been incorrect in categorising part-time farming as a transitory state between rural and fully urban occupations.[39] As a mode of production, part-time farms exemplify a permanent feature of the restructuring of regional rural economies. Two types of part-time farmer can be identified. In one group are farmers with a second business; they are as likely to be ex-urbanites moving to the country as farm-families taking up urban employment. In the second and more numerous group are 'worker-peasants' (Arbeiterbauer) who commute each day for factory and office employment in nearby urban areas. In peripheral, often mountainous regions such farmers have taken up a miscellany of employment including forestry, public services and tourism. Nevertheless, part-time farming is seldom found in areas with large farm structures and remains a 'survival strategy' for the occupiers of small holdings.

Over 44 per cent of farmers in the EC can be classified as part-time, and Figure 16B describes the development of part-time farming by the late 1970s. West Germany is the principal focus of development; 54 per cent of farmers have 'other employment', a feature that stems from deliberate government policy to site industry in rural areas. High proportions of part-time farmers are also found in southern parts of Italy, including Lazio, Umbria and Trentino-Alto Adige, Belgium, and Roussillon-Languedoc in southern France. By comparison, part-time farming is relatively undeveloped elsewhere.

Agricultural policy makers hold an ambivalent attitude towards part-time farming. On the one hand, off-farm employment helps to sustain farms and rural services while supplementing otherwise low farm incomes on small holdings. When linked to tourism, it can be a means of fostering economic development, for example in peripheral regions with mountain environments. On the other hand, off-farm employment also slows down the process of farm-size change, especially in peri-urban locations, and perpetuates a stratum of small farms with their social farm-income problems. There is only limited evidence, however, of a 'blocking' effect on structural change in upland environments.[40] In addition, part-time farms are prone to over-mechanisation, the idling of land, and lower productivity compared with their full-time equivalents. The role of part-time farming in rural development, therefore, is still contested[41] and is not specifically recognised by the CAP.

Conclusion

The large number of small farms with less than ten hectares of land constitute the central problem faced by the CAP. Although considerable reductions have been made in the number of farms and the size of the workforce in the last three decades, the pace of change has not been sufficient to resolve the problem of low farm incomes in the Community. Nevertheless, the farm-size/farm-income problem varies within the EC (Figure 14B) and is more severe throughout Italy, Mediterranean France, and central and south-western parts of West Germany than in most regions of Belgium, the Netherlands and the United Kingdom.

The failure of the structural policies of the CAP can be attributed to three factors. First, the rate of structural change in agriculture is determined by employment opportunites elsewhere in the economy. The economic recession of the late 1970s and 1980s severely limited such opportunities and lay beyond the scope of agricultural policy. Secondly, there has been divergent economic development both amongst Member States and between regions within each country. Unfortunately, countries and regions with the most pressing farm-structure problems have tended to perform poorly in general economic terms. Thirdly, policy makers have failed to appreciate the magnitude and obstacles against structural change. Indeed, their policy decisions on price levels, financial aid to modernise farms, and direct income supports to LFAs all obstruct rather than assist structural change in agriculture. Here lies one of the many contradictions of agricultural policy in the EC as in other developed countries.

Unfortunately, the CAP has exacerbated rather than diminished regional variations in average farm income. Direct and indirect price support under the Guarantee Section has favoured northern regions of the Community in direct proportion to gross output per farm. Payments from the Guidance Section, together with national aid, have been insufficient to alter materially the relative status of the 'rich' and 'poor' agricultural regions of the EC. Thus price policy has been an indirect and, therefore, an inappropriate way of approaching the income problems of small farms and specific regions. Moreover a general relationship exists between high agricultural incomes and high general incomes in a region.[42] Even here there is evidence of regional divergence in the EC. The ratio between the overall GDP of the ten weakest and the ten strongest regions of the Community widened from 1:2.9 in 1970 to 1:4.0 in 1977. The problem of divergent agricultural incomes, therefore, is part of a general problem of the EC as an economic system. There now exists a widespread recognition of the need for comprehensive programmes of regional development to counteract the divergent economic trends between the 'core' and 'periphery' of the Community.[43] This, however, leads into a discussion of policy options for the CAP which is a main concern of the following chapter.

Notes

1 The terms 'farm' and 'holding' are not strictly synonymous since a farm can be comprised of several holdings. A holding is the unit of land occupied by a person making the return to an agricultural census. When a person occupies several holdings an arbitrary choice can often be exercised on how many separate census returns are made. For convenience, however, the two terms are used interchangeably in this book.

2 The processes which create spatial variations in farm size are discussed in most text-books on agricultural geography or regional geography. For example, for France they are described by J. Tuppen, *The economic geography of France* (London, Croom Helm, 1983), pp. 53–63.

3 J. Scully, 'The evolution of farm structures in the Community', in M. Tracy and I. Hodac, *Prospects for agriculture in the EC* (Bruges, De Tempel, 1979), pp. 139–59.

4 R. Calmès, 'L'évolution des structures d'exploitation dans les pays de la C.E.E.', *Annales de Géographie* XC (1981), pp. 401–27.

5 M. Tracy, *Agriculture in Western Europe: challenge and response 1880–1980* (2nd ed.), (London, Granada, 1982), p. 325.

6 G. Weinschenck and J. Kemper, 'Agricultural policies and their regional impact in Western Europe', *European Review of Agricultural Economics* VIII (1981), pp. 251–81.

7 D. K. Britton and B. Hill, *Size and efficiency in farming* (Farnborough, Saxon House, 1975).

8 G. P. Hirsch and A. H. Maunder, *Farm amalgamation in Western Europe*

(Farnborough, Saxon House, 1978); OECD, *Structural reform measures in agriculture*, (Paris, OECD, 1972).

9 Levels of payment are varied from time to time to take account of price inflation.

10 D. Boulet, 'Les co-opératives et leur environment socio-économique', *Revues des études co-opératives* CLXIX (1972), pp. 339–64.

11 J.-C. Lebossé, 'Croissance des co-opératives et developpement régional', *Norois* CXII (1981), pp. 465–81.

12 G. Foxall, *Co-operative marketing in European agriculture* (Aldershot, Gower, 1982).

13 J. Gilbank, 'La modernisation de la viticulture française', *Geographia Polonica* XXIX (1974), pp. 260–70.

14 R. E. Williams, 'Milk marketing in a European framework', *Journal of Agricultural Economics* XXXI (1980), pp. 311–20.

15 I. R. Bowler, 'Structural change', in M. Pacione (ed.), *Progress in rural geography* (London, Croom Helm, 1983), pp. 46–73.

16 R. L. King and S. Burton, 'Structural change in agriculture: the geography of land consolidation', *Progress in Human Geography* VII (1983), pp. 471–501.

17 A recent comparison of farm and non-farm incomes in the United Kingdom has been made by J. K. Bowers and P. Cheshire, *Agriculture, the countryside and land use* (London, Methuen, 1983), pp. 82–3. Their estimates are subject to all the limitations listed on p. 201 and in addition do not cover the period of deteriorating farm incomes in the late 1970s and early 1980s. Nevertheless, farm incomes in the United Kingdom are shown as increasing to a level comparable with a number of professional socio-economic groups and above those of male manual workers. This relatively favourable farm income situation – also repeated in Belgium and the Netherlands – is confirmed by a comparison of the proportion of the workforce employed in agriculture and the proportion of GDP generated by agriculture (Table 1). An approximate equality between farm and non-farm income levels in aggregate is indicated when the two proportions are equal. Inequalities in income levels by farm size, farm type and region are masked by these data.

18 *Agra Europe*, August 27th, 1982, p. E/2.

19 P. Henry, *Study of the regional impact of the Common Agricultural Policy*, Regional Policy Series 21, RICAP Working Group (Brussels, Commission of the European Communities, 1981), pp. 145–53.

20 This methodological approach has been attempted by Henry, *Regional impact of the Common Agricultural Policy* (1981), pp. 24–32.

21 G. Podbielski, 'The Common Agricultural Policy and the Mezzogiorno', *Journal of Common Market Studies* XIX (1981), pp. 332–50.

22 M. Benedictis, 'Agricultural problems in Italy: national problems in a Community framework', *Journal of Agricultural Economics* XXXII (1981), pp. 275–86.

23 Scully, *Prospects for agriculture in the EC* (1979), p. 153.

24 Little research has been published on the use and impact of the 'Mediterranean Package' of measures and they are not considered further here.

25 G. Caffarelli, *L'agriculture française et la politique agricole commune* (Paris, Chambre d'Agriculture, 1982), p. 21.

26 This theme has been researched for the United Kingdom by I. R. Bowler, 'Adoption of grant aid in agriculture', *Transactions of the Institute of British Geographers*, New Series (1976), pp. 143–58.

27 Anon, 'Failure of EEC's farm structure policy?' *Agra Europe* CMXV (1981), p. E/1.

28 Arkleton Trust, *Schemes of assistance to farmers in less favoured areas of the EEC*, (Langholm, The Trust 1982), p. 5.

29 W. Peters and U. Langendorf, 'Direct income transfers for the agricultural sector in Less Favoured Areas', *European Review of Agricultural Economics* VIII (1981), pp. 41–55.

30 M. MacEwen and G. Sinclair, *New life for the hills* (London, Council for National Parks, 1983), p. 10.

31 Peters and Langendorf, *European Review of Agricultural Economics* VIII (1981), pp. 41–55.

32 A–M. Discamps-Dheur, 'Les conséquences régionales de la politique agricole commune', *Revue Economique du Sud-Ouest* IV (1980), pp. 57–69.

33 J–C. Bontron and N. Mathieu, 'Transformations agricoles et transformations rurales en France depuis 1980', *Economie Rurale* CXXXVII (1980), pp. 3–9.

34 This book is not concerned with wider societal changes in rural areas as described by M. Gervaise *et al*, *Une France sans Paysans* (Paris, Editions du Seuil, 1965) and M. Debatisse, *La Revolution silencieuse, le combat des Paysans*, (Paris, Calmann Lévy, 1963), nor with more recent trends in rural repopulation such as 'population turnaround' and 'counter-urbanisation'.

35 C. Canevet, 'De la polyculture paysans à l'intégration: les couches sociales dans l'agriculture', *Norois* XXVI (1979), pp. 507–22; S. H. Franklin, *Rural Societies*, (London, Macmillan, 1971).

36 A. Bourgeois and M. Sebillotte, 'Reflexion sur l'évolution contemporaine des exploitations agricoles', *Economie Rurale* CXXVI (1978), pp. 17–28; Centre for European Agricultural Studies, *The future of the family farm in Europe*, Seminar Paper 1, CEAS, (Wye College, University of London, 1974).

37 M. Phlipponneau, 'Politique agricole et problèmes régionaux', *Problèmes Economiques* CLXXXVII (1970), pp. 24–8; Williams, *Journal of Agricultural Economics* XXXI (1980), pp. 311–20.

38 R. Taylor, 'The Galway farmers', *New Society* XXXII (1975), pp. 405–8.

39 Centre for European Agricultural Studies, *Part-time farming: its nature and implications*, Seminar Paper 2, CEAS, (Wye College, University of London, 1976); OECD, *Part-time farming in OECD countries* (Paris, OECD, 1978); W. Frank, 'Part-time farming, underemployment and double activity of farmers in

the EEC', *Sociologia Ruralis* XXIII (1983), pp. 20–27.

40 J–C. Jauneau, *La pluriactivité des agricultures de montagne*, Etude 124, Groupement de Grenoble, (CTGREF, Ministère de l'Agriculture, 1978).

41 R. Gasson (ed.) *The place of part-time farming in rural and regional development*, Seminar Paper 3 (CEAS, Wye College, University of London, 1977).

42 Henry, *Regional impact of the Common Agricultural Policy* (1981), p. 72.

43 Scully, *Prospects for agriculture in the EC* (1979), p. 159.

11
Conclusion – the second enlargement and prospects for the CAP

This final chapter draws together the themes and conclusions of the earlier parts of the book. The issue linking together all the material is the changing geography of agriculture within the EC and the role of the Common Agricultural Policy in those changes. In addition, the themes and conclusions have been related to the second enlargement of the Community and the prospects for the CAP. This approach is particularly apposite at the time of writing when the future of the CAP is more uncertain than for many years.

Conclusions

The development and application of agricultural policy in the context of the EC has coincided with a period of rapid modernisation of farming in Western Europe. This macro-economic trend owes little to the CAP. Farm modernisation was underway before the Community was founded and similar trends have been documented in many other countries with developed market economies. Farm modernisation has been associated with a technological revolution in the production methods used in agriculture, the substitution of capital inputs for farm labour, the evolution of a large-farm structure at the expense of small production units, and the specialisation in production of farms and agricultural regions. The penetration of the farming system by both the food-processing sector and non-farm capital has also characterised recent developments in agriculture although these two features have fallen outside the scope of this book.[1] All of these trends have evolved at different rates amongst the countries and regions comprising the EC, and variations in the agricultural geography of the Community have been heightened rather than diminished. Overall, agriculture has expanded its capacity to produce food at a faster rate than the increase in domestic demand.

Nevertheless, the CAP, together with its wider institutional setting, has played an important role in influencing the scale and speed of farm modernisation in the EC, and not always with beneficial results. By setting stable institutional prices above those prevailing on the world market (Table 43), the CAP has provided a relatively risk-free environment in which investment for farm modernisation has been encouraged. The speed and scale at which new farming technologies have been applied owe much to the favourable price situation engineered by agricultural policy in the EC. Unfortunately, the CAP has failed to develop an environmental concern to counteract the damaging consequences of farm modernisation for the conservation of countryside flora, fauna and landscapes. In the United Kingdom, in particular, concern has been voiced about the threat to moorlands, wetlands, hedgerows and woodlands posed by an expanding arable area.[2] The scale and speed at which the threat has grown appears proportional to the prosperity created by contemporary CAP price levels and by grants and subsidies to modernise individual farms.

The CAP, through the intervention system, has also provided an assured market for a wide range of products. With the discipline of the market removed, producers have been able to exand their production beyond the domestic needs of the Community. Developed in the 1950s, the intervention system was not designed to handle the structural surpluses of agricultural products that emerged in the 1970s. Moreover, the lower-priced international market has been unable to absorb such surpluses without the payment of expensive export subsidies by the EC. At fault is not so much the intervention system itself, but the failure of policy makers to evolve further measures to deal with the changing circumstances of a modernised agricultural industry.

An assured market has allowed the trend to product specialisation and concentration by farm, region and country to proceed at a faster rate than might otherwise have occurred. Further specialisation has been possible even in products surplus to domestic demand. Unfortunately, specialisation tends to exacerbate rigidity in production patterns. Increasing difficulty is encountered in reorientating production in line with market demand when individual farms and whole regional farming systems are committed to the production of a limited number of products.

At the same time, favourable prices and an assured market have slowed down the rate of structural adjustment in agriculture, arguably to socially acceptable levels. Consequently, the EC retains a large number of very small farms but they are still not economically viable even under contemporary price-cost conditions. Small farms tend to be concentrated in a few regions of the Community and so perpetuate spatial variations in farm incomes. Moreover, institutional prices are generally set at a level sufficient to maintain the viability of smaller family farms. In countries with a predominantly large-farm structure, such as the United Kingdom, aggregate farm incomes can reach excessively favourable levels compared with non-farm incomes.[3]

The slow rate of structural change in agriculture also reflects the failure of the Community's structural programmes under the CAP. The steep rise in the cost of implementing price supports has prevented the EC from devoting resources to such programmes, and their development has been left largely to national governments. Structural reform, while proceeding, has not been politically acceptable in the Member States at the high rate required to resolve the farm-income/surplus production conflict. Nor does the 'structural', or 'resource adjustment', solution to the farm problem now go unchallenged. In practice, enlarged farms, like their small counterparts, have to increase output to maintain profit levels. The resources available to large farms allow them to increase production at a faster rate than on small farms, and a change in farm-size structure does not of itself, therefore, resolve the problem of over-production. Indeed so great is the scope for modernisation after farm enlargement in Western Europe that the potential for even higher yields is enormous. The problem will be resolved only by an emphasis on the economic use of resources in production, rather than the volume of resources, but this aspect of agricultural policy receives little emphasis under the present structure of the CAP. Nor is the superior economic efficiency of the large farm now accepted without question. A number of recent studies have shown that the major gains from scale economies occur in expanding from a one-to a two-man, or medium-sized farm, and thereafter there are few economies with increasing scale of production.[4]

Another facet of the challenge to the 'structural solution' of agricultural problems has been created by the projected, long-term rise in the price of energy, and the emphasis laid on the finite resources available for the production of petroleum. Large farms are critically dependent on petroleum products for both motive power for machinery and the chemical base for fertilisers and pesticides. In the long term, an agricultural industry based on energy-conserving family farms, rather than very large, capitalistic farms may well offer a more secure basis for food production. This last view is held as yet by a minority, but it is significant that Sicco

Mansholt, once the arch-proponent of the 'structural solution', now sees great merit in it.[5]

Very variable product prices have evolved under the CAP. Countries have had to make varying adjustments by product to new common price levels: conversion (green) rates for CAP prices from units of account into national currency equivalents have varied by country, supported by monetary compensatory amounts; the degree of market security and price protection has varied by product. Despite intensive research, no simple, clear-cut relationships have been established between these price variations and production trends for various products under the CAP. This conclusion applies to aggregate production trends in the EC, as well as to changes in the location of production whether viewed at the national or regional levels of analysis. While agriculture has undergone spatial variation in intensification, concentration and specialisation, factors other than CAP prices have been influential. In particular, there have been national and regional changes in production costs stemming from national variations in price inflation and the general level of economic development in the economy. Also, national schemes of assistance to the agricultural sector, including non-tariff barriers to trade, have been used to shelter regional farming systems from full competitive trading within the common market. Inter-regional competition in agriculture, therefore, has varied with changes both in product prices and input costs, and overall has tended to intensify rather than alter pre-existing patterns of production within the EC. The assumption that the market has allocated production efficiently between regions is in some doubt. Regional comparative economic advantages appear to have been modified by separate national governmental decisions on acceptable inflation and currency exchange rates, and on the interpretation and implementation of directives under the CAP.

The overall impact of the CAP, therefore, involves a balance between benefits and costs even though they cannot be quantified directly. On the side of the benefits can be placed the stimulation given to the modernisation of farming and the increased productivity of agriculture. In addition, the security and continuity of food supply has been assured for consumers, with benefits for the stability of the domestic market, while the rate and scale of change in farming communities has been held to socially acceptable levels. Unfortunately, these benefits appear to be outweighed by the costs incurred by the CAP.

First, the production of structural surpluses in commodities such as milk, sugar and cereals has not been resolved with consequences for the balance of expenditure from FEOGA (Table 44). Secondly, the budgetary cost of the CAP has escalated, almost seventy per cent of expenditure being committed to buying, storing and trading food surpluses. For a period in the late 1970s, expenditure from FEOGA increased by nearly 23 per cent a year while the growth of 'own resources' remained at 12 per cent a year. Consequently, by 1983 the limits of the resources available to the Budget had been reached thus creating a financial crisis in the Community. Thirdly, the CAP exerts a detrimental impact on the pattern of world trade in agricultural products, particularly where export subsidies are used to dispose of surplus production. Fourthly, price levels and price relativities between products remain out of line with market demand so that consumer prices are higher than they need be. This creates an indirect, hidden cost for the consumer that in some countries is larger than the direct costs of the CAP (Table 6). In addition, the welfare costs of higher food prices are regressive, since poorer families expend a higher proportion of their income on food than richer families. The benefits of the CAP pass disproportionately to large farms which produce most of the food in the EC, and an effective transfer of income takes place from poorer, mainly urban families to richer farm families. The social equity, if not morality of such a situation is questionable. Fifthly, within agriculture the disparity in income levels between large and small farms, and between rich and poor regions has been widened rather than diminished by the CAP. Direct and indirect price support under FEOGA passes to the producers of milk, cereals and sugar beet, all of which are located in northern rather than southern parts of the EC (Figure 7A). Moreover, in only a few countries (Belgium, the Netherlands, United Kingdom) have farm and non-farm incomes come into an approximate balance; elsewhere in the Community farm incomes remain low and actually fell in real terms in several years in the early 1980s. Sixthly, the modernisation of farming has created external costs in terms of environmental damage but these are difficult to quantify and, outside conservation interests, are too often viewed as the inevitable price of farming progress.

The argument now deployed is that the excessively high costs of the CAP do not stem from the policy measures themselves. Rather, as agriculture has become modernised, policy makers have failed to allow the CAP to evolve in needed directions and have not manipulated existing measures effectively. A central issue has been the Community's decision-making structure. This allows interest groups to exert an influence disproportionate to their size. National farming unions, when organised in a federation at the EC-level, have proved adept at exploiting the complex checks and balances that are a feature of decision-making in the Community. They have been able to maintain favourable price levels for producers and thwart attempts to develop policy measures that curb the production of agricultural surpluses. By comparison, the countervailing interests of consumers and taxpayers have been poorly organised, and even the European Parliament has been unable to develop a concensus of opinion on how best to operate and develop the CAP. Farm interests, therefore, have been able to keep the gains they have achieved, and this inertia of possession has been termed the 'ratchet principle' of public support.[6] Only the recent, direct involvement of the Finance Ministers in establishing the annual Budget of the EC appears to be a force sufficient to counteract the otherwise powerful farming interest.

The continued primacy of national, rather than Community interests in decision making within the Council of Ministers has also worked against the evolution of a more successful agricultural policy. Decisions have continued to be short-term rather than long-term in perspective, and have favoured inertia and incrementalism through failure to agree on policy proposals. Sufficient of the Member States have gained a national benefit from the existing structure and operation of the CAP, together with the method of funding FEOGA, as to counter proposals to limit price increases, curb production, or alter the regional disbursement of support. The allocation of payments from FEOGA by product, for example, tends to favour countries such as France, West Germany, Italy and the Netherlands at the expense of countries such as the United Kingdom, Ireland and Denmark (Table 51). The disbursement of payments reflects the national concentration of production of each product (Table 15) and the extent to which surplus commodities are exported or held in intervention stores. Even countries such as West Germany, which carry a large proportion of the direct costs of the CAP, have not consistently opposed the inertia of the CAP. Internal political considerations, the effective national lobbying of Agriculture Ministers by farming interests, and countervailing benefits in the industrial sector[7] have been the main causes.

There has also been a broadly-based failure in political will to confront many of the problems created by inertia in the CAP. Rather, politicians have focussed their attention on the direct costs of the CAP and, latterly, on the equity of net national contributions to the Budget. This has diverted public and political attention away from arguably more important issues such as the magnitude and incidence of the indirect welfare costs of the CAP, the impact of EC policies on world trade, the environmental costs of policy measures, and the growing inequalities in regional farm incomes. A number of recent

Table 51. Payments from the Guarantee Section (FEOGA) to Member States, 1980 (% total payments for each product)

Product	F	WG	It	N	Country[a] UK	D	B	Ir	L	
Milk	21	27	0	23	8	8	6	6	0.2	(100)
Cereals (and rice)	43	19	14	5	8 '	3	7	0.1	0	(100)
Sugar	43	23	6	7	6	3	11	1.2	0	(100)
Oils and fats	20	20	49	1	9	0.7	0.6	0	0	(100)
Beef and pigmeat	20	22	15	10	8	8	2	16	0	(100)
Fruit and vegetables	14	4	79	2	0.3	0.1	1	0	0	(100)
Total payments	25	22	16	14	8	5	5	5	0.1	(100)
Total agricultural output[b]	28	19	21	9	14	4	3	2	0.1	(100)

a: country abbreviations as on Table 39; b: by value
Source: G. Caffarelli, *L'agriculture française et la politique agricole commune*, (Paris, Chambre d'Agriculture, 1982), p. 7

books, academic papers and television programmes have exposed the wider costs of the CAP,[8] but the major political parties in the EC appear unwilling to confront these issues.

A major cause of this unwillingness lies in the 'cornerstone' role of the CAP within the Community. The viability of the EC tends to be judged, rightly or wrongly, by the success of its economic policies. Since the CAP is the most developed of any policy, its continued existence, including a supra-national character, is vital to the credibility of the EC as a political entity. The prominence of the CAP is maintained by the failure of a concensus to emerge in the development of other common policies. This failure can be attributed to continuing national differences in economic philosophy on the role of the state, differences in economic and social structures, and differences in economic performance.[9] Consequently, the CAP is locked into considerations of global 'high' politics, with the EC perceived as defending the democratic societies of Western Europe in an increasingly hostile international, political and economic environment.[10] The CAP, despite its many contraditions and problems, is defended in its present form for other than strictly economic or agricultural reasons. In this perspective, national interests converge on the wider international political benefits of maintaining the CAP and outweigh divergent national interests over the narrow economic costs of farm policies. Nevertheless, it seems unlikely that the CAP will be used again as an agent for further integration within the Community. Federalist and neo-functionalist views of integration command little support from either politicians or the public, not least because a transfer of loyalties to the EC has not taken place amongst either bureaucrats or politicians, nor has there been a separation of politics and economics in decision making. The nation state, which has proved more resilient and persistent than thought possible in the 1950s when the EC was being created, is likely to remain on balance a force for divergence rather than convergence in the Community.[11]

A final reason for the continuing problems of the CAP lies in the failure of policy makers to take sufficient account of the agricultural geography of the EC. The agricultural sector has tended to be treated as if it were homogeneous in character. Several chapters of this book have been devoted to examining the spatial variations that exist between and within the Member States in features such as farm size, farm incomes, the location of production of various crops and livestock, and the process of farm modernisation. When product prices and other policy measures are institutionalised but fail to take account of spatial variations in agriculture, regional farm-income differentials can be deepened and unwanted production can be stimulated in countries or regions unduly favoured by the pricing system. In addition, divergent national and regional trends in the development of agriculture have come to reflect divergences in the wider national and regional economies. The CAP, while evolving a regional dimension, has not yet taken sufficient account of the reality of spatial variations in the agricultural sector of the EC.

The second enlargement
The first enlargement of the European Community took place in 1973 and extended membership to Denmark, Ireland and the United Kingdom. The second enlargement, involving Greece, Portugal and Spain (hereafter termed 'The Three'), is being spread over a longer period of time beginning in 1981 with the accession of Greece to the Treaty of Rome. Portugal and Spain, having applied for membership in 1977, still have their applications under consideration at the time of writing. While the second enlargement has implications for all aspects of the economic, social and political life of the Community, attention here is turned to those features that have a bearing upon agriculture.

Decision making
Some of the forces which prompted the initial formation of the EC have also been influential in the second enlargement. They comprise a mixture of 'high politics' and national economic self-interest (see Chapter 1). On the one hand, the existing members of the Community wish to strengthen the

southern flanks of the NATO alliance against a perceived political and military threat from Eastern Europe. A second enlargement would protect three newly-emerged democracies and carry forward the wider geo-political goal of a united Western Europe.[12] On the other hand, Greece, Portugal and Spain are hopeful of finding it easier to deal with their economic, social and political difficulties in the Community framework.[13] The expected economic benefits for agriculture include increased exports to members of the EC, higher farm prices and access to export subsidies, raised farm incomes, access to the structural and marketing grants of the CAP, and a net inflow of aid under FEOGA.

The political decision by the Council of Ministers to accept an enlargement of the Community preceded the Commission's economic analysis of the potential consequences.[14] For once, the Commission rather than the Council was responsible for raising a cautionary voice in the affairs of the Community. The economic analysis, broadly supported by subsequent academic research, stressed how existing problems of the EC would be thrown into relief and given urgency by the second enlargement. In agriculture, for example, it is not easy to discern the benefits to the existing members but they might include the increased export of temperate agricultural products that are currently in surplus, together with the inflow of cheaper Mediterranean products onto northern markets. Faced by the mounting evidence of the costs of a second enlargement, many Member States have become less enthusiastic and have resisted progress in enlarging the Community. Italy and France, for example, have questioned whether Portugal and Spain have converged sufficiently on the EC in terms of their economic and social structures to make membership desirable. They have pressed for a long (ten-year) transition period in any further enlargement. Other countries, such as West Germany and the United Kingdom, have linked revisions of the CAP and the Budget to agreement on the second enlargement. Given the decision-making process of the Community, the lack of concensus in these two areas has effectively blocked progress on the applications by Portugal and Spain. Moreover, their applications have been linked together, and the major problems posed by the Spanish economy have spilled-over to effect the less problematic application from Portugal.

Greece, Portugal and Spain share a number of characteristics that differentiate them from other Member States. They have a generally lower level of economic development, a higher dependence on agriculture for employment and GDP, and regionally concentrated economic development. In so far as they share a 'Mediterranean identity', a second enlargement would promote a change in the balance of political power within the Community. 'Southern' rather than 'northern' interests would be given greater weight in decision making, but only Italy of The Nine would gain from such a change. The increased heterogeneity of the EC would increase conflicts between national interests, and even in France the political weight of the Mediterranean *départements* could not be ignored. Two developments are possible. Decision making in the Council of Ministers could become based more regularly on the principle of majority voting, but with a clearer procedure laid down for identifying issues which fall under the veto of a specific national interest.[15] Alternatively the Community could divide into a two-tier structure, with a looser form of membership for those countries unable to accept the supra-national character of the EC. This two-tier structure has already been rejected as leading to a weakening of the Community as a political entity. Without some change, however, the second enlargement appears to be a prescription for further obstruction in the process of integration, with minimalist tendencies in the development of policies such as the CAP. Without some agreed long-term objective for the EC, countries will remain unwilling to ignore the balance of advantage and disadvantage of belonging to the Community.[16]

Production structures

The failure of France and Italy to press the cases of Portugal and Spain for membership stems in part from problems in the agricultural sector. Agriculture remains an important element in the economy and society of these countries, more so than in Italy which, apart from Greece, is the most agrarian country in the Community (Table 52). Taken together, The Three would increase the number of farms in the EC by 57 per cent, agricultural workers by 55 per cent, the agricultural area by 49 per cent, and total agricultural production by 24 per cent.[17] The sheer aggregate size of agriculture in The Three poses problems for its assimilation in the EC.

Moreover, the products of The Three compete directly with the agricultural output of the Mediterranean regions of Italy and France (Table 52), and less directly with the horticultural produce of Belgium and the Netherlands. Crops predominate and include citrus fruit, vegetables, olive oil, wine, rice and tobacco. Such crops comprise 80 per cent of the agricultural output of Languedoc-Roussillon and 56 per cent of that of southern Italy. In addition, there is a danger of creating surpluses in wine, olive oil, certain fresh and processed fruit and vegetables (Table 53). Portugal, for example, has an existing surplus of wine and olive oil; Spain has surpluses in citrus fruit (oranges, grapefruit), olive oil, wine and rice; Greece

Table 52. Agricultural production structures of The Three (% total agricultural output by value)

Product	Spain	Portugal	Greece	(Southern Italy)
Meat	26	23	18	14
Vegetables	15	10	10	16
Other vegetable products	12	15	21	11
Fruit and citrus fruit	12	13	12	17
Cereals	10	12	13	11
Milk	9	8	7	5
Wine	6	10	3	11
Olive oil	5	6	10	12
Other animal products	5	3	6	3
:Total livestock	:40	:34	:31	:22
:Total crops	:60	:66	:69	:78
Agriculture as % GDP	7.5	12.8	16.1	8.0[a]
Agriculture as % workforce	18.9	29.0	27.3	12.8[a]
% farms below 5 ha	57.5	77.3	72.5	68.5[a]
% lands in farms below 5 ha	10.8	14.9	38.9	21.6[a]

a: Italy
Source: M. Tracy and I. Hodac, *Prospects for agriculture in the EEC*, (Bruges, De Tempel, 1979), p. 218; J. B. Donges *et al.*, *The second enlargement of the EC* (Tübingen, J. C. B. Mohr, 1982), pp. 26, 32, 146

is similarly placed for citrus fruit, wine, sugar, peaches and rice. Greece alone raised the EC's self-sufficiency in olive oil to 95 per cent, and with the inclusion of Spain and Portugal this figure would be raised to a surplus. One estimate has placed the possible surplus at 200,000 tonnes a year by 1990 in an EC of twelve members.

Too much, however, should not be made of the similarities between The Three. They differ considerably in economic size and economic structure, in their potential for increases in agricultural output, and in their previous economic ties with the EC. Greece, for example, had twenty years of Association status before joining the Community and was able to harmonise many aspects of agricultural policy over an extended period. Moreover, Greece's relatively small agricultural sector was absorbed relatively easily into the EC. Thus the five-year transition period negotiated by Greece (seven years for a few agricultural products) is not readily transferable to the other applicant countries. Spain in particular is judged to have an enormous potential for increased agricultural production, although the provision of irrigation is a critical limiting factor. Certainly, the rate of increase in agricultural output since 1960 has been higher in Spain than in the Member

States. Spain's potential for generating surpluses in Mediterranean products such as olive oil, wine, certain fruits (peaches) and vegetables is a major problem for the second enlargement (Table 53). Portugal, by comparison, has the least dynamic agricultural sector of The Three, and during the last decade has suffered falling production in half of its major products – olive oil, wheat, barley, apples and tomatoes. The political and economic upheavals of the country are usually blamed for the falling productivity of Portuguese agriculture.

Just what production trends will occur in Greece and would occur in Spain and Portugal under the existing CAP is difficult to predict. As has been stressed in earlier chapters, CAP and even market prices are a poor guide to production trends in agriculture. Of equal importance are changing production costs, national rates of inflation and currency exchange rates. Even so, CAP prices tend to be above those on the domestic markets of The Three. In Portugal, for instance, CAP prices would be substantially higher for wheat, rye, rice, olive oil and sugar, lower for pigs, and approximately the same for beef and veal. There is little doubt that higher prices under the CAP would stimulate production in The Three but to an unknown extent and in a way that would vary regionally within each new Member State. Overall, citrus fruit, olive oil, wine and certain vegetables seem most likely to expand in production under the CAP.

Table 53. Levels of self-sufficiency in The Three (%)

Product	Greece	Portugal	Spain	The Three	The Twelve
Wheat	104	30	96	96	103
Rice	138	65	120	120	80
Sugar	118	4	90	90	112
Potatoes	107	97	103	103	100
Wine	126	129	122	122	103
Olive oil	112	97	125	125	103
Peaches	205	97	104	104	109
Vegetables	106	105	111	111	98
Citrus fruit	179	100	243	247	78
Beef and veal	48	71	87	74	95
Pigmeat	90	97	97	96	100
Poultrymeat	101	100	99	100	103
Sheepmeat	100	100	100	100	74
Butter	70	71	90	81	115
SMP	0	67	22	27	104
Eggs	101	100	105	104	101

Source: J. B. Donges *et al.*, *The second enlargement of the EC* (Tübingen, J. C. B. Mohr, 1982), p. 148

The Mediterranean regions of southern France and Italy would suffer most from such production trends.[18] Moreover, with the increasing regional specialisation of production under the CAP, the capacity of these regions to reorientate their production has been limited even further. A dependence on wine, fruit and vegetables, in association with a small-farm structure, renders these regions vulnerable to competition from The Three. One estimate for wine production showed labour costs per hour in Spain to be lower by 47 per cent compared with southern France.[19] Thus Roussillon would suffer competition from Spanish peaches and apricots, Provence competition from pears and table grapes, and both regions competition from tomatoes and early potatoes. Unfortunately these regions would not have a strong competitive advantage in the markets of The Three in products such as milk, butter, SMP, beef/veal, maize and sugar. The scope for increased competitive trading by the existing members of the Community would be taken up mainly by producers in northern regions. A second enlargement, therefore, would exacerbate the income problems already experienced by the Mediterranean regions of the EC.

Farm-size structure

There is broad agreement that the second enlargement would magnify existing regional economic inbalances within the Community. The proportion of the workforce employed in agriculture would be raised, but the proportion of GDP generated by farming would not be increased by the same amount. Regional income differences would be deepened by the small farm-size structure of Portugal and Greece – over 70 per cent of farm units in these countries occupy less than five hectares of agricultural land. Not only is the average size of farm small, but the degree of land fragmentation and concentration is more severe than elsewhere in the Community. Farm holdings in Spain are comprised on average by 10.8 separate plots; comparable figures for Greece and Portugal are 6.5 and 6.4 respectively. For land concentration, 7 per cent of Spain's farms (over fifty hectares) occupy 68 per cent of the farmland. Of course strong regional variations exist in these features despite government programmes of farm consolidation and land reform.[20] In Portugal, for example, the *minifundia* of the north west contrast with the *latifundia* of the south east. Taking a broad view, the ratio of the value of agricultural product per worker between rich and poor regions, which fell from 74 per cent in The Six to 70 per cent in The Nine, is expected to deteriorate further to 58 per cent in an EC of The Twelve.[21]

The lack of industrialisation in The Three, and the weak growth of the economies of all West European countries, poses further problems. Few alternative employment opportunities exist for the large numbers of farmers and farm workers expected to be displaced from agriculture in The Three over the next decade. The problem extends to the long-distance migrants of southern Europe who traditionally have found employment in prosperous northern industrial regions. Given the inadequacy of the CAP as a way of redistributing incomes, the need for broadly-based programmes of integrated regional development is now widely accepted. Such programmes are expected to co-ordinate regional, national and Community initiatives in areas such as soil conservation, afforestation, water control for irrigation, infrastructure (roads, electricity), small businesses (farm implements, food processing), fisheries, and service industries (tourism). The costs of regional development programmes for The Three would be enormous. One proposal in 1983, just for the south of France, Italy and Greece, recommended the investment of 6628 million ECU (£4043 million) spread over six years in an 'integrated Mediterranean programme'.[22]

Trade

There is little doubt that trade flows would alter in an enlarged Community. The Three are already important trading partners of the EC in agricultural products, and that trade can be expected to increase. In 1980, for example, the Community purchased 51 per cent of Greece's total food exports, 59 per cent of Spain's, and 49 per cent of Portugal's. The degree of dependence on the EC varies by product. Over 90 per cent of Spain's exports of oranges, tomatoes, and potatoes, for example, go to the Community, but only 38 per cent of exports of olive oil and 34 per cent of wine. Preferential trade agreements have allowed The Three to make these inroads into the markets of the EC, especially in fruit and wine. The share of the Italian market taken by The Three, for example, has steadily increased in fruit and wine, but has fallen in vegetables, tobacco and olive oil. In the French market, The Three have gained a larger share of trade in fruit and olive oil.[23] Such competition has lead Italy and France to press for a long transition period should Spain, in particular, be accepted as a member of the Community.

Other members of the Community should not think that a second enlargement would materially resolve the problems of surpluses in 'northern' products. This is despite Spain and Portugal, unlike Greece, being net food importers and drawing relatively little of their present food supplies from the EC. Total present-day imports of meat to The Three represent less than 2 per cent of the Community's annual production; the proportions for cereals and butter are seven and eleven per cent respectively.[24] In addition, The Three are as likely to increase their own production of meat and cereals under CAP prices and assured markets as other members of the EC. Thus existing surpluses of meat and animal products would shrink in an enlarged Community, but not to the extent that the problem of surpluses would be resolved. Indeed new surpluses in products such as wine, olive oil and citrus fruit would be created.

Imports from other non-member Mediterranean states would be displaced by the second enlargement. Once again trade in citrus fruit, wine and vegetables would be most affected. Countries like Cyprus, Morocco and Israel, with high proportions of 'sensitive' agricultural products in their food exports to the EC, would be treated adversely. Morocco, which sends 96 per cent of its tomato exports to the EC, would suffer a loss of such trade together with potatoes, wine and olive oil. Similarly Tunisia, with 87 per cent of its olive oil exports being purchased by the EC, would also lose trade. In addition, should Mediterranean crops from the EC begin to attract export subsidies, other Mediterranean countries would meet increased competition on world markets.

Not all non-member countries would suffer. The ACP states, for example, would enjoy an enlarged market for the export of their tropical products. Equally, the protection against non-member countries is actually higher at present in Spain and Portugal than would be afforded by the CET. The applicant countries would also have to accept the EC's preferential trading agreements with other Mediterranean countries. The balance of advantage and disadvantage in trade, therefore, is as complex for non-member states as it is for the applicant countries and members of the EC.

Funding the CAP

The issue that most concerns 'northern' members of the Community is the impact of the second enlargement on the Budget. Most estimates show that together The Three would increase net expenditure once price supports under the CAP, payments from the Guidance section, and programmes under the Social and Regional Funds are accounted for.[25] Given the small farm-size structure of The Three, for instance, Guidance payments from FEOGA could well be doubled. Again, with the CAP unchanged, a projected surplus of 200,000 tonnes of olive oil would attract finance of approximately £1 billion a year. These demands would exceed the already precarious level of 'own resources'. A second enlargement, therefore, could not be contemplated without some revision of the present method of funding the Budget. Either spending on agriculture would have to be reduced, or else increased 'own resources' would have to be approved. Perhaps both developments would be necessary.

Taken together, these observations suggest a delayed rather than early decision on the further enlargement of the EC. Arguably, the Community at present is not sufficiently robust to withstand the increased strains of a second enlargement, and in any event prior decisions on revising the CAP and the Budget are needed. The economic costs of an enlargement currently appear to outweigh the political benefits.

Prospects for the CAP

The CAP faces a number of pressures for change: farm incomes (protection), political (unity), budgetary (control) and external trade (respectability).[26] There are no easy options to resolve any of these pressures and the CAP has been subject to a succession of proposals for revision both from the Commission and agricultural economists.[27] For convenience, the numerous and often contradictory proposals have been simplified and gathered under three headings: options for reform, evolution and devolution.[28]

Reform options

The term 'reform' describes adjustments to the CAP within the existing framework of its principles and objectives, and is the usual approach adopted by the European Commission. The reform options fit the political realities of the EC better than either the evolution or devolution options, although arguably they are less able to resolve the pressures for change in the long term. Indeed, it would be remarkable if a framework for agricultural policy developed in the late 1950s and early 1960s could continue to be adapted sufficiently to meet the very changed circumstances of a modernised agriculture in the 1980s. Even so, none of the proposed options is without its problems and costs, and this contributes to the present impasse in the evolution of the CAP.

The reform gaining most widespread support is a progressive reduction of CAP price supports nearer to levels applying in the EC's major competing countries. The aim would be to reassert the discipline of the market so as to reduce costly surpluses in products such as milk, cereals and sugar. A firm long-term price signal would be provided for farmers to induce them to alter the intensity and balance of their production. Proposals also exist for the co-responsibility levy, at present limited to sugar and dairy production, to be extended to other farm products. Producers would participate more fully in financing the disposal of their surplus production. In addition, budget liability could be limited by the wider imposition of a super-levy – a tax on sales above a certain quota as introduced for dairy products in 1984 – together with an extension of the existing system of production thresholds. More extreme quantity limits could be placed on the volume of produce to be taken into intervention; excess production would have to seek a return on the open market.

Substantial commodity, country and welfare effects emanate from price reductions with a varied impact on producers, consumers, and taxpayers. The methodological framework for comparing the varied costs and benefits is still being developed,[29] but it is clear that the net overall gains to society from a policy change are nearly always less than the losses to at least one party. The incidence of the loss, however, varies according to the type and level of the price change.

Other recent policy options advanced by the Commission include 'reviewing' food imports into the EC together with a more 'positive' export policy. The damaging consequences of both options for world trade need little elaboration. Moreover other traditional, alternative policy measures offer few firm promises of greater success. Production quotas, for example, as operated for sugar in the EC, have the effect of fossilising existing patterns of production even though a proportion of a national quota may be set aside for new entrants. Competition between producers and the specialisation of production in favoured regions is stifled. There are also political difficulties in allocating 'rights to produce' between Member countries together with problems of administrative complexity when allocating quotas to individual farms. Consumer subsidies are often advanced as an alternative method of clearing the market of surplus products, but again there are difficulties. Because price elasticities of demand are low for most products, the price reduction necessary to clear the market can be very large and the budgetary costs equally so. In addition, a new stratum of complicated marketing mechanisms becomes necessary, and overall the burden of farm support passes from consumers to taxpayers. Deficiency payments,[30] as operated in the United Kingdom prior to 1972, also offer few possibilities for the CAP. The visibility of payments made to producers would be high, while the EC has a limited ability to raise the taxes necessary to operate a deficiency payments system.

Clearly the occupiers of small farms, and producers in the LFAs, would suffer a damaging loss of income under a sustained reduction in the level of price supports. Proposals to extend direct income payments to small holdings in problem regions have been favourably evaluated,[31] but support for them by the Commission has not included any details on possible levels of assistance.

Another facet of the reform option lies in the area of conservation. The CAP could develop explicit objectives for conservation in matters such as wildlife, flora, and landscape.[32] Lowering the profitability of agriculture through reduced price levels would reduce the economic incentives both for extending agriculture into environmentally sensitive areas such as moorlands, woodlands and wetlands, and for intensifying agricultural output. In the short-run, however, farmers could well intensify their production in the face of falling prices so as to protect their net farm incomes. This is sometimes described as a 'perverse supply-response'. In the long term, however, production would fall in line with prices and so reduce the contemporary pressure placed on the environment by relatively high agricultural support prices. Another option would be the extension of planning controls over changes in the rural landscape.[33] Moorland ploughing, wetland drainage, hedgerow removal, woodland felling, and new farm buildings would all require planning approval. The farming lobby would clearly oppose this option, while its effectiveness at the Community level would depend on the existing planning framework in each Member State. Part of this planning-control option would be the regulation of the use of herbicides, pesticides and fungicides in agriculture. A final option for conservation lies in persuading the farming community to be more active in practicing conservation. A growing body of opinion points up the need to reinforce persuasion by withdrawing grants and subsidies which finance ecologically damaging practices such as hedgerow removal. Two recent proposals have developed an alternative package of agricultural subsidies and incentives.[34] This approach seeks to link financial expenditure on approved conservation practices in farming to limits on the types of grant aid which intensify agricultural output.

Evolution options

The view can be taken that as vital as the reform options are, they will not succeed by themselves in resolving the pressures for change in the CAP. In a modernised, capitalistic agriculture the 'liberal alternative' becomes increasingly feasible.[35] This involves a clear separation of the price and income roles of the CAP. The option requires a progressive but relatively rapid withdrawal of intervention on domestic markets, relying for support on protection at the common frontier. Smaller farmers would be compensated throughout the EC by direct income payments. These could be reduced in step with the increasing business size of the farm and weighted according to the severity of the handicaps to production in mountain and hill regions.[36] The high direct cost of this option would probably require some national funding by each Member State. In addition, upper limits would be needed on the density of livestock or the volume of crop production per hectare eligible for support so as to prevent the further intensification of production.

A number of criticisms are commonly levelled against direct income payments. Farm organisations, for example, assert that farmers resent income supplements. Experience under the Less Favoured Areas Directive does not lend support to this view, while the criticism often appears to be a defence of the status quo with its selectively favourable price supports. On the other hand, direct income supplements do appear to remove incentives to improve the efficiency of farming. On this matter there is no reason why receipt of payments should not be conditional upon the implementation of an approved farm plan, as under Directive 72/159 (see Chapter 10). Producers would gain the benefits of reduced production costs. Another problem would undoubtedly lie in the increased bureaucratic machinery necessary to make payments to millions of farmers. However, existing channels for making social security payments could be used and would make clear the need to separate price and income measures under the CAP.

Alternatively a discriminatory price system could be introduced.[37] On each farm, the price for small quantities of production would be high but would fall per unit of output with increasing quantities of sales. There would probably be a contraction of output by the largest producers and such a system would clearly discriminate against countries and regions with a large-farm structure. However, the price system could be adapted to reflect the farm-size structure of each type of product with advantages for the occupiers of small farms in the low-income, Mediterranean regions of the EC.

Neither of these options really faces the issue that while small farms are problematic, they display marked regional concentrations. Moreover, supporting the incomes of such farms offers no long-term solution to the regional farm-income problem. Consequently, perhaps the main evolutionary policy option for the CAP is its more direct involvement in regional development policy. There are two feasible developments. Price incentives could be altered from region to region to take account of the changing needs of different groups of farmers. In some regions, for example, prices might be used to promote a greater diversification of production away from products in surplus. In other regions, the output of a particular product might be stimulated to bring economies of scale to the regional food processing industries. A second development would involve the transfer of resources from the Guarantee to the Guidance Section of FEOGA. A regionally-oriented policy would vary the application of structural measures according to the particular needs of each region. On occasions part-time farming would be directly encouraged in parallel with investments in non-farm employment under the Regional Development Fund. Elsewhere, the continued out-migration of farm-families might have to be financed or regional co-operative marketing structures strengthened. An integrated regional development option has already been developed under the CAP. The case here is for an extension and strengthening of that option in other problem rural regions of the EC with an integration into the work of the Regional Development Fund.[38] Experience under the CAP shows that farm income problems will not be resolved by the reform options alone; the CAP is inadequate as a residistributive mechanism. This role is best filled by the explicitly selective financial assistance of the Regional Development Fund.

Devolution options

The effects of national self-interest in decision making within the EC have been documented throughout this book. Increasingly policy proposals are being advanced whereby this reality is faced. At issue is a partial devolution or re-nationalisation of agricultural policy. The continuation, if not extension of MCAs, for instance, has been supported on the grounds that national price differences can be maintained.[39] Extending this concept, trade between the Member States could take place at price levels agreed by the Council of Ministers (common trading prices), together with import levies and export restitutions under a common financial responsibility.[40] National prices to domestic producers, however, could be set at levels deemed desirable by each country and which could reflect national variations in farm size and production structures. Each country would be responsible for funding its own level of domestic prices. Some commentators observe that this system would do no more than is accomplished under common funding at present by the system of MCAs.

There is some accord between the devolution option and a central theme of this book, namely that the CAP at present fails to take sufficient account of the varying geography of agriculture in the EC. In an ideal world, pricing would be based on agricultural regions defined other than by national boundaries, and prices would make allowances for regional variations in the location of production and farm-size structures. The treating of agriculture as if it were a homogeneous entity limits the effectiveness of most agricultural policy measures. A further advantage of the devolution option is that with a return to partial national funding, pressure is removed from the 'own resources' of the CAP. This would be helpful in the discussions on the second enlargement, for each country would be responsible for funding its own agricultural surpluses. Of course the devolution option fails to remove the conflicting interests of net importing and net exporting countries in setting the level of common trading prices. In addition it strikes at the principles underlying the CAP. Whether the devolution option would actually bring about the demise of the EC itself is by no means clear, but it is a gamble Member States are not prepared to take. It would certainly mean an end to the CAP as it was first established and would put back progress in the process of integration.

The most likely outcome for the CAP is a package of measures drawn from the reform, evolution and devolution options. Of these, reform measures are likely to predominate in a further process of gradual adaptation to the pressures for change. No once-and-for-all solution to the problems of the CAP is possible, nor does speculation on a 'last chance for change' take account of the resilience of the CAP. The most likely scenario is a CAP characterised by a recurring state of crisis, but with a long-term trend away from price policies towards direct income payments. Whether the Member States can accept substantial and continuing transfers of resources to regions in other countries within the EC will determine the progress made in developing the regional dimension of the CAP. In the author's view, only a regional dimension gives a long-term prospect for resolving the agricultural problems of the Community.

Notes

1 Further discussion of the food processing industry can be found in J. Burns *et al* (eds.), *The food industry – economics and policies* (London, Heinemann, 1983).

2 G. P. Wibberley, 'Strong agricultures but weak rural economies – the undue emphasis on agriculture in European rural development', *European Review of Agricultural Economics* VIII (1981), pp. 155–70.

3 This theme is developed by J. K. Bowers and P. Cheshire, *Agriculture, the countryside and land use* (London, Methuen, 1983).

4 D. K. Britton and B. Hill, *Size and efficiency in farming* (Farnborough, Saxon House, 1975).

5 Soil Association, *The Common Agricultural Policy* (Stowmarket, The Association, 1979).

6 Bowers and Cheshire, *Agriculture, the countryside and land use* (1983), p. 66.

7 R. Cecil, *The development of agriculture in Germany and the United Kingdom. Vol. 1: German agriculture 1870–1970* (Centre for European Agricultural Studies, Wye College, University of London, 1979), p. 67.

8 For example – R. Body, *Agriculture: the triumph and the shame* (London, Temple Smith, 1982).

9 U. Everling, 'Possibilities and limits of European integration', *Journal of Common*

Market Studies XVIII (1980), pp. 217–18.

10 S. Hoffman, 'Reflections on the Nation-State in Western Europe today', *Journal of Common Market Studies* XXI (1982), pp. 21–37.

11 Y–S. Hu, *Europe under stress* (London, Butterworths, 1981).

12 A. Fernandez, 'L'agriculture Espagnole et la Communauté Économique Européenne', *Economie Rurale* CXXIII (1978), pp. 31–5.

13. G. Gallus, 'Agricultural problems of the accession of Greece, Portugal and Spain to the EC', *Intereconomics* 1 (1979), pp. 6–10.

14 W. Wallace and I. Herreman (eds.), *A Community of Twelve? The impact of further enlargement on the European Community* (Bruges, De Tempel, 1978), p. 419.

15 Wallace and Herreman, *A Community of Twelve?* (1978), pp. 12–13.

16 J. B. Donges *et al*, *The second enlargement of the European Community* (Tübingen, J.C.B. Mohr, 1982), p. 13.

17 Commission of the European Communities, *Enlargement of the Community – general considerations*, Bulletin of the European Communities 1/78 (Brussels, The Commission, 1978), p. 9.

18 A–M. Discamps-Dheur *et al*, *L'élargissement de la Communauté Européenne*, Documentation interne de la politique regionale dans la Communauté 12 (Brussels, Commission of the European Communities, 1981).

19 M. Tracy and I. Hodac (eds.), *Prospects for agriculture in the EEC* (Bruges, De Tempel, 1979), p. 208.

20 M. Guedes, 'Recent agricultural land policy in Spain', *Oxford Agrarian Studies* X (1981), pp. 26–43.

21 Tracy and Hodac, *Prospects for agriculture in the EEC* (1979), p. 211.

22 Commission of the European Communities, *The Commission Proposals for the integrated Mediterranean programmes*, COM (83) 24, (Brussels, The Commission, 1983); Commission of the European Communities, *Enlargement of the Community – economic and sectoral aspects*, Bulletin of the European Communities 3/78 (Brussels, The Commission, 1978), p. 23.

23 Tracy and Hodac, *Prospects for agriculture in the EEC* (1979), pp. 218–19.

24 Wallace and Herreman, *A Community of Twelve?* (1978), p. 204.

25 Commission of the European Communities, *Enlargement of the Community*, 3/78 (1978), p. 41; Tracy and Hodac, *Prospects for agriculture in the EEC* (1979), p. 215.

26 T. E. Josling *et al*, *Options for farm policy in the EC* (London, Trade Policy Research Centre, 1981), pp. 7–25.

27 A useful survey for the early 1970s has been made by R. Fennell, *The Common Agricultural Policy: a synthesis of opinion*, Report 1 (Centre for European Agricultural Studies, Wye College, University of London, 1973).

28 Similar options for restructuring the Budget have not been discussed here but can be found in G. Denton *et al*, *Reform of the Common Agricultural Policy and*

re-structuring of the EEC Budget (London, University Association for Contemporary European Studies, 1983), pp. 18–32.

29 A. E. Buckwell *et al*, *The costs of the Common Agricultural Policy* (London, Croom Helm, 1982), p. 90.

30 These are direct payments to producers based on the difference betweeen the market price and a previously agreed 'guaranteed' price.

31 U. Koester and S. Tangermann, 'Supplementing farm policy by direct income payments', *European Review of Agricultural Economics* IV (1977), pp. 7–31.

32 Bowers and Cheshire, *Agriculture, the countryside and landuse* (1983), pp. 133–59.

33 M. Shoard, *The theft of the countryside* (London, Temple-Smith, 1981).

34 C. Potter, *Investing in rural harmony* (London, World Wildlife Fund, 1983); M. MacEwen and G. Sinclair, *New life for the hills* (London, Council for National Parks, 1983), pp. 33–7.

35 G. Weinschenck and J. Kemper, 'Agricultural policies and their regional impact in Western Europe', *European Review of Agricultural Economics* VIII (1981), pp. 251–81.

36 MacEwen and Sinclair, *New life for the hills* (1983), pp. 43–4.

37 J. S. Marsh, 'Proposals and prospects for further CAP reforms', in Denton *et al*, *Reform of the Common Agricultural Policy* (1983), pp. 56–77.

38 P. S. T. McGee, *Development of the Regional Policy of the European Communities*, Working Paper 65 (Oxford, Oxford Polytechnic, Department of Town Planning, 1982).

39 T. Heidhues *et al*, *Common prices and Europe's farm policy* (London, Trade Policy Research Centre, 1978). In 1984, however, the Council of Ministers agreed to phase out MCA payments.

40 J. S. Marsh, 'European agricultural policy: a federalist solution', A New Federalist Paper, Reprinted from *New Europe* (Winter 1976/7).

Appendix
GLOSSARY OF ACRONYMS

ACA(s)	Accession Compensatory Amount(s)
ACP	African, Caribbean, and Pacific (countries)
AHGS	Agriculture and Horticulture Grants Scheme
ASSILEC	*Association de l'Industrie Laitière de la Communauté Européenne*
AUA(s)	Agricultural Unit(s) of Account
AWU(s)	Annual Work Unit(s)
Benelux	Customs union comprising Belgium, the Netherlands and Luxembourg
BEUC	Bureau Européen des Unions le Consommateurs
CAP	Common Agricultural Policy
CCA	Commercial Co-operation Agreement
CCC	Consumers' Consultative Committee
CET	Common External Tariff
CIAA	Commission des Industries Agricoles et Alimentaires
CIBE	Confederation of European Sugar Beet Producers
CNJA	*Centre National des Jeunes Agriculteurs*
COGECA	*Comité Général de la Co-opération Agricole*
COMECON	Council for Mutual Economic Assistance
COMEPRA	*Comité Européenne pour le Progrès Agricole*
COPA	*Comité des Organisations Professionnelles Agricoles*
COREPER	*Comité des Représentants Permanents*
DBV	*Deutsche Bauernverband*
EC	European Community
ECSC	European Coal and Steel Community
ECU(s)	European Currency Unit(s)
EEC	European Economic Community
EFTA	European Free Trade Association
EMS	European Monetary System
EPD	European Progressive Democrats (group in the European Parliament)
EPP	European People's Party (group in the European Parliament)
ESC	Economic and Social Committee (of the Commission)

EUA(s)	European Unit(s) of Account
Euratom	European Atomic Energy Community
EURO-COOP	European Community of Consumer Co-operatives
FAC	Food Aid Convention
FDP	*Freie Demokratische Partei Deutschland*
FEOGA	*Fonds Européen d'Orientation et de Garantie Agricole* (also referred to as the Agriculture Fund)
FHDS	Farm and Horticulture Development Scheme
FNPL	*Fédération Nationale des Producteurs de Lait*
GATT	General Agreement on Tariffs and Trade
GDP	Gross Domestic Product
GSP	Generalised System of Preferences
GVA	Gross Value Added
HGCA	Home-Grown Cereals Authority
HLCA	Hill Livestock Compensatory Allowances
IBAP	Intervention Board for Agricultural Produce
IMF	International Monetary Fund
IWA	International Wheat Agreement
LFA(s)	Less Favoured Area(s)
MCA(s)	Monetary Compensatory Amount(s)
MEP(s)	Members of the European Parliament
MLC	Meat and Livestock Commission
NATO	North Atlantic Treaty Organisation
NFU	National Farmers' Union
OEEC	Organisation for European Economic Co-operation
ONIC	*Office National Interprofessionnelle des Céréales*
SAFER	*Sociétés d'Aménagement Foncier et d'Établissement Rural*
SCA	Special Committee for Agriculture
SPD	*Social Demokratische Partei Deutschland*
SIDO	*Société Interprofessionnelle des Oléagineux*
SMP	Skimmed Milk Powder
Stabex	Stabilisation of Export Earnings (Scheme)
UA	Unit of Account
UNCTAD	United Nations Conference on Trade and Development
VAT	Value Added Tax
VIB	*Voedselvoorzieningsin-en Verkoopbureau*
WFP	World Food Programme

Index